The Future of Fusion Energy

The Future of Fusion Energy

Jason Parisi
University of Oxford, UK

Justin Ball
Swiss Federal Institute of Technology in
Lausanne (EPFL), Switzerland

NEW JERSEY · LONDON · SINGAPORE · BEIJING · SHANGHAI · HONG KONG · TAIPEI · CHENNAI · TOKYO

Published by

World Scientific Publishing Europe Ltd.
57 Shelton Street, Covent Garden, London WC2H 9HE
Head office: 5 Toh Tuck Link, Singapore 596224
USA office: 27 Warren Street, Suite 401-402, Hackensack, NJ 07601

Library of Congress Cataloging-in-Publication Data
Names: Parisi, Jason, author. | Ball, Justin (Justin Richard), author.
Title: The future of fusion energy / by Jason Parisi (University of Oxford, United Kingdom),
　　Justin Ball (Swiss Federal Institute of Technology in Lausanne (EPFL), Switzerland).
Description: New Jersey : World Scientific, 2018. |
　　Includes bibliographical references and index.
Identifiers: LCCN 2018016230 | ISBN 9781786345424 (hc : alk. paper)
Subjects: LCSH: Nuclear fusion. | Renewable energy sources. | Power resources.
Classification: LCC QC791 .P37 2018 | DDC 539.7/64--dc23
LC record available at https://lccn.loc.gov/2018016230

British Library Cataloguing-in-Publication Data
A catalogue record for this book is available from the British Library.

First published 2019 (Hardcover)
Reprinted 2019 (in paperback edition)
ISBN 978-1-78634-749-7 (pbk)

Copyright © 2019 by World Scientific Publishing Europe Ltd.

All rights reserved. This book, or parts thereof, may not be reproduced in any form or by any means, electronic or mechanical, including photocopying, recording or any information storage and retrieval system now known or to be invented, without written permission from the Publisher.

For photocopying of material in this volume, please pay a copying fee through the Copyright Clearance Center, Inc., 222 Rosewood Drive, Danvers, MA 01923, USA. In this case permission to photocopy is not required from the publisher.

For any available supplementary material, please visit
http://www.worldscientific.com/worldscibooks/10.1142/Q0160#t=suppl

Desk Editors: Dr. Sree Meenakshi Sajani/Jennifer Brough/Shi Ying Koe

Typeset by Stallion Press
Email: enquiries@stallionpress.com

Printed in Singapore

To fusioneers — past, present, and future

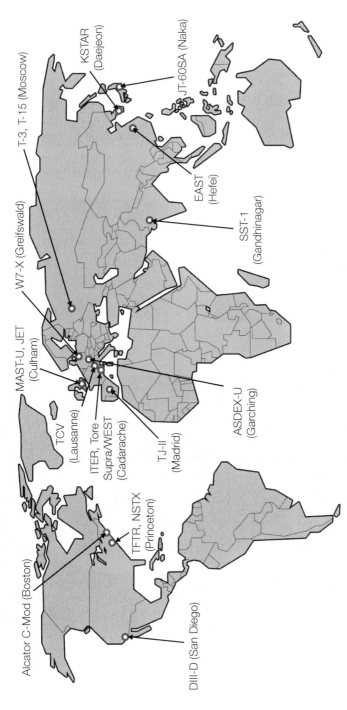

The locations of some of the world's most significant magnetic fusion devices.

(a) TFTR, which used to be in Princeton, USA, produced 10 MW of fusion power.

(b) Alcator C-Mod was in Boston, USA and explored high magnetic field operation.

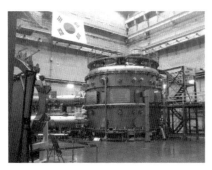

(c) KSTAR is a superconducting tokamak in Daejeon, South Korea.

(d) NSTX, in Princeton, USA, is a spherical tokamak, which means it has a small donut hole.

(e) JET, the world's largest tokamak, is near Oxford, UK and produced 16 MW of fusion power.

Some of the world's most significant magnetic fusion devices.

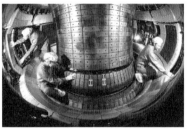

(f) EAST, in Hefei, China, was the world's first fully superconducting tokamak.

(g) ASDEX-U, in Munich, Germany, discovered an improved way to operate tokamaks: H-mode.

(h) JT-60SA is under construction in Naka, Japan and will soon be the world's largest fusion device.

(i) Wendelstein 7-X is the world's largest stellarator and began operation in Greifswald, Germany in 2016.

(j) WEST explores the behavior of long duration tokamak operation in Cadarache, France.

More of the world's most significant magnetic fusion devices.

Preface

As two young scientists researching nuclear fusion, we see a massive disparity between the current state of the field and the descriptions available to the public. Our goal in writing this book is to bridge that gap to provide you with the basics of fusion and the ability to judge its future prospects for yourself. This is not a textbook, but it still explains matters in sufficient technical detail to reveal the logic that shapes research and development. For the most part, this can be done without mathematics, which we hope will make this book accessible to a multitude of audiences: scientists, policymakers, and laypeople alike. Basically, anyone who wants to learn more about the triumphs and challenges of fusion. However, we do not claim that understanding nuclear fusion is easy — fusion has yet to deliver electricity to the grid precisely because there are so many complicated and counterintuitive things to understand! We hope that you will stick with it and have some fun in the process.

Despite popular conception, fusion science and technology has made remarkable progress, even compared to other fast-moving fields. It is not a pie-in-the-sky dream, but a multi-faceted scientific and technological challenge that will yield to sufficient brainpower. Yet, we acknowledge that there is a long way to go before fusion will be providing commercial electricity to the grid. This book will clarify the main sticking points.

As a practical note, throughout this book you will find "Tech Boxes" that serve as brief asides to the main text. The book is written to be understandable without them, but they are a good starting point if you want more technical details on a particular topic. Additionally, there are several short "Basics Boxes" that cover fundamental material such as units and mathematical notation. If you already know this material, then feel free to skip them.

Finally, in a popular science book on a field as extensive as fusion energy, we could not fit in every important topic. There are many fascinating areas that we have entirely skipped or left grossly over-simplified. This is particularly true for the history of fusion research as there are already several excellent popular works available. Instead, we have set our sights forward, focusing on how current technology and the laws of physics motivate the global effort to develop a commercial fusion power plant. Enjoy!

Jason and Justin
Oxford, UK
April 2018

About the Authors

Justin Ball is a post-doctoral researcher at the Swiss Federal Institute of Technology in Lausanne, where his research focuses on plasma shaping in tokamaks. He holds a Masters in Nuclear Engineering from MIT and a PhD in Theoretical Physics from the University of Oxford. Previously, he has worked at the Perry Nuclear Power Plant, Bettis Atomic Power Laboratory, and Lawerence Livermore National Laboratory.

Jason Parisi is a PhD student at the University of Oxford, where his research focuses on turbulence in the edge of fusion plasmas. Jason has previously done plasma physics research at Princeton Plasma Physics Laboratory, Tsinghua University, and the University of Science and Technology of China. Jason obtained his undergraduate degree from Yale University.

@JB_Fusion
@JFP_Fusioneer

Acknowledgments

As with fusion research in general, this book was an international collaboration — not simply a product made by its authors. A large number of talented and kind people from all over the world enabled it, both directly and indirectly. To this end, we would like to thank Felix Parra, Stephan Brunner, Ian Hutchinson, Rob Goldston, Michael Barnes, Alex Schekochihin, Anne White, Steve Cowley, Colin Roach, Dennis Whyte, Greg Hammett, and Peter Catto for providing the authors with a world-class education. We are also grateful to Rob Goldston, Bill Spears, Mark Henderson, Marcelo Baquero-Ruiz, Paola Paruta, Ambrogio Fasoli, Holger Reimerdes, Brandon Sorbom, Kim Fuellenbach, Pedro Molina Cabrera, Stephan Brunner, Anna Teplukhina, and Daniel Kammen for reading through various chapters and spotting ALL of our embarrassing mistakes. We are particularly indebted to Aaron Scheinberg and Hugo de Oliveira for taking the book we had written and, through their comprehensive feedback, developing it into what you see today. Lastly, thanks to the entire Ball family for asking questions until our unintelligible technobabble became understandable English.

We would also like to thank our wonderful editors at World Scientific for their vision, skill, dedication, and patience: Jennifer Brough, Elena Nash, Sree Meenakshi Sajani, and Chandrima Maitra.

On a personal note, Justin would be remiss if he did not acknowledge the EUROfusion Consortium and the funding he has received from the Euratom research and training program 2014–2018 under Grant Agreement No. 633053. He is also grateful to the Swiss Plasma Center for being a nice place to work. Finally, he loves his family and their constant, unwavering encouragement.

Jason would like to thank all his amazing friends and family for their support. Most of all, Spence Weinreich. He is also indebted to the University of Oxford's Theoretical Physics Department and all the brilliant people who fill it, for making the PhD experience so rewarding. Finally, he would like to thank the Bulletin of the Atomic Scientists for the Leonard M. Rieser Fellowship, which is where this book project began.

Contents

Preface   xi

About the Authors   xiii

Acknowledgments   xv

Introduction: The Case for Fusion   xxiii

PART 1   MOTIVATION   1

1. The Hydrogen-Powered Civilization   3

    1.1   Revolutions in Energy Use 3
    1.2   Comparing Options 7

2. Energy in Numbers and Graphs   11

    2.1   Can We Even Consume Energy? 11
    2.2   A Brief History of Energy 12
    2.3   Our Energy Resources 17
         2.3.1   Fusion 19
         2.3.2   Nuclear fission 23
         2.3.3   Geothermal 27
         2.3.4   Solar 28
         2.3.5   Wind 30
         2.3.6   Biomass 35
         2.3.7   Fossil fuels 37
         2.3.8   Hydroelectric 39

	2.3.9	Tidal	41
	2.3.10	Wave	44
2.4	Tackling Intermittency		46
	2.4.1	Energy storage	50
	2.4.2	Demand management	54
	2.4.3	Expanding electrical grids	54
	2.4.4	Extra generating capacity	56
2.5	What is "Renewable"?		57
2.6	Outlook		58

PART 2   THE BASICS   63

3. Fundamentals of Fusion Energy   65

3.1	The Nuclear Potential	65
3.2	Binding Energy	70
3.3	Fusion Cross-Section	77
3.4	Fusion Fuels	83
3.5	Plasma	86

4. Plasma Confinement   89

4.1	Quantifying Confinement	90
4.2	Magnetic Fields	91
4.3	Electric Fields	94
4.4	Electrostatic Confinement	96
4.5	Linear Magnetic Confinement	100
4.6	Combing a Hairy Ball	103
4.7	Particle Drifts	104
4.8	Toroidal Magnetic Confinement	109
4.9	Magnetic Surfaces	114
4.10	Bananas and Super-Bananas	118
4.11	MHD Stability	120
4.12	Classical and Neoclassical Transport	123
4.13	Turbulent Transport	126
4.14	The Lawson Criterion and the Triple Product	131
4.15	Where is Magnetic Fusion Now?	136

5. Fusion Technology — 139

- 5.1 Magnets 139
- 5.2 Plasma Heating and Current Drive 144
  - 5.2.1 Inductive 145
  - 5.2.2 Neutral beam 148
  - 5.2.3 Electromagnetic wave 150
- 5.3 First Wall 153
- 5.4 Divertors 157
- 5.5 Tritium Breeding Blanket 160
- 5.6 Vacuum Vessel 164
- 5.7 Diagnostics 164
- 5.8 Radioactive Waste and Remote Maintenance 167
- 5.9 Generating Net Electricity 169

PART 3   THE STATE OF THE ART — 173

6. The Past: Fusion Breakthroughs — 175

- 6.1 1920s: Understanding Stars 175
- 6.2 1950s: A Kick-Start for Fusion 179
- 6.3 1960s: Superconducting Magnets 181
- 6.4 1960s: The Tokamak 186
- 6.5 1970s: Bootstrap Current 189
- 6.6 1980s: H-Mode 193
- 6.7 1980s: Plasma Shaping 198
- 6.8 1990s: Deuterium–Tritium Fuel 202
- 6.9 2000s: Supercomputers 204

7. The Present: ITER — 211

- 7.1 ITER's Goals 212
- 7.2 ITER's Strategy 216
  - 7.2.1 Heating systems 218
  - 7.2.2 Divertor 220
  - 7.2.3 First wall 222

7.3   ITER's Schedule and Cost 224
7.4   Transition to DEMO 232
7.5   Other Things to be Excited for 234

8. The Future: Designing a Tokamak Power Plant   237

8.1   Power Plant Design from First Principles 238
8.2   Maximizing Net Electric Power 241
8.3   Maximizing Plasma Pressure 243
8.4   Maximizing Plasma Current 246
8.5   Maximizing Magnetic Field Strength 247
8.6   Minimizing External Power 249
8.7   Minimizing Heating Power 249
8.8   Maximizing Plasma Density 253
8.9   Minimizing Current Drive Power 253
8.10 Maximizing Material Survivability 255
8.11 Striking the Right Balance 257

PART 4   SPECIAL TOPICS   261

9. Alternative Approaches to Fusion Energy   263

9.1   Stellarators 263
9.2   Inertial Confinement Fusion 269
9.3   Private Fusion Startups 277
      9.3.1   Tokamak Energy Ltd. 281
      9.3.2   General Fusion 284
      9.3.3   Lockheed Martin 287
      9.3.4   TAE Technologies 290
      9.3.5   Lawrenceville Plasma Physics 296
      9.3.6   Helion Energy 300
      9.3.7   Commonwealth Fusion Systems 301

10. Fusion and Nuclear Proliferation   303

10.1 Nuclear Physics: A Double-edged Sword 310
10.2 Building Nukes 312
      10.2.1 Uranium enrichment 317

		10.2.2 Plutonium production 320
		10.2.3 Weapon designs 321
	10.3	Conventional Fission Reactors 325
	10.4	Breeder Reactors 330
	10.5	Fission Proliferation Risks 333
	10.6	Fusion Proliferation Risks 336
	10.7	The Nuclear Energy Transition 341
	10.8	Reshaping Geopolitics 343
	10.9	Being a Role Model 344

11. Fusion and Space Exploration   347

    11.1 Basics of Spaceflight 349
    11.2 Fusion Thruster 354

PART 5   CONCLUSIONS   357

12. When Will We Have Fusion?   359

Bibliography   367

Index   371

Introduction: The Case for Fusion

Fusion powers the universe. Every one of the stars in the sky uses fusion to generate enormous amounts of energy. Why shouldn't we? Since the end of World War II, scientists, engineers, and policymakers have worked to emulate nature and use fusion to solve our energy problems.

Nuclear fusion, once industrialized, will tremendously increase the amount of energy available to humans. As the world's needs continue to grow, fusion could provide an invaluable source of low-carbon electricity and, in the long run, would be able to supply the majority of our energy.

This book consists of 12 chapters, all of which are intended to be accessible to a reader with no formal physics background. Some parts may be particularly challenging (e.g. Chapters 3, 4, and 8), but it is our hope that the reader sticks with these demanding areas, as they will ultimately be the most useful. Below, we have sketched a brief summary of our key arguments, both to serve as a reference and a guide for what lies ahead.

- **Fusion is an attractive, sustainable solution to humanity's energy problems.** Fusion fuel has the highest energy density of any demonstrated source, meaning it tends to be cheap to gather and transport. Moreover, compared to renewable sources like wind or solar, a fusion power plant would take up minimal area. There are practically limitless fuel reserves for fusion here on Earth and elsewhere in the universe. These reserves are readily

- **Fusion is a massively complex problem that requires significant upfront investment.** For a society to get to the point where it can build a fusion power plant is nontrivial. It requires the construction of increasingly intricate and expensive experimental devices, an industrial capability to engineer and manufacture complex components, and the political will to sustain a massive long-term research effort.
- **Given current funding levels, we should NOT count on fusion to be our primary low-carbon electricity source for the 21st century.** Without a large increase in research and development, fusion will not be a panacea for tackling the problems of climate change. It does not look like it will be economically feasible quickly enough. The world needs to be building low-carbon energy generation now. Nevertheless, fusion could very well generate most of our electricity by 2100 and provide a long-term solution to our climate woes.
- **Our current ignorance and the complexity of fusion are reasons for confidence in fusion.** Power plant design is a balancing act between many different competing considerations from many different fields of engineering and physics. It is not an elegant, unified problem. Rather, it is a complex web of interacting considerations compounded by the limitations of our understanding. As we learn more, we can tailor the design of a power plant in order to use progress in one field to compensate for difficulties in another. Hence, a failure to achieve fusion would require long-term stagnation in many disparate fields.
- **ITER is crucial.** ITER, the world's biggest science experiment, is currently being built by an international collaboration of 35 countries. As such, the world's fusion research is focused on supporting this single machine. While ITER is expected to demonstrate the scientific and technological feasibility of fusion, its failure could set the community back by decades. Nevertheless, ITER will undoubtedly make seminal contributions to the knowledge of

humanity and could pave the way for a demonstration power plant. Unfortunately, it will not be completed until at least 2025.
- **Small and simple is a must.** After we have learned from experiments on ITER, it is important to research strategies to enable smaller and simpler fusion devices. Such devices can be built quickly and cheaply, allowing one to iterate through design ideas faster and rapidly leverage the results of ITER into a commercial power plant. More importantly, it reduces the risks associated with trying new things, allowing fusion experiments to be more daring.
- **Fusion can mitigate nuclear proliferation concerns, strengthening nuclear security.** Simply put, it would be very difficult to quickly and secretly use fusion reactors to generate fissile material for nuclear weapons. By replacing conventional nuclear fission power plants with fusion, the world can eliminate the need for enriched uranium and plutonium, making nuclear bombs much more difficult to produce.
- **Fusion has been underfunded for decades.** The current funding for fusion is peanuts. In 2016, the US Department of Agriculture projected the cost of peanut subsidies to be nearly double the entire budget of the Office of Fusion Energy Sciences. This isn't to say that peanuts aren't important, but that relatively little is spent on fusion. Hence, a dramatic acceleration in the development of fusion could be achieved with a small shift in public support and/or political priorities.
- **Fusion research brings substantial benefits to many fields, from basic science to engineering.** The development of fusion will continue to stimulate research in a wide range of fields, from superconducting technology to material science, computer simulations to applied mathematics.
- **Fusion brings the world together.** During the height of the Cold War, fusion energy research was one of the best examples of US–USSR collaboration. The international nature of fusion has only grown stronger since. Working together to develop technology strengthens ties between countries and is a strong driver of global peace and stability.

- **First-generation fusion reactors will be large power plants, but over time smaller models may become available.** There are fundamental physical reasons why first-generation reactors will be large. However, improvements in our physics understanding and technological advancements might enable much smaller reactors. This means that, while first-generation devices will be limited to grid electricity generation, advanced fusion reactors may be small enough to power ships and spacecraft.[1]
- **A fusion propulsion system could open up the Solar System (and beyond) to human exploration and settlement.** Because of their high exhaust velocities, fusion rocket engines can be very efficient. They could be capable of taking humans to Mars in as little as 20 days, to Titan (the largest moon of Saturn) in 150 days, and our neighboring star systems in 75 years.

This is a condensed version of the next several hundred pages. We wish to paint a realistic picture of the potential for fusion energy and accurately convey the scale of the difficulties that must be overcome. However, we would like to emphasize that *fusion is a realizable energy source that, if given sufficient investment, can provide watts to the grid.* In this book, you will be guided through the fundamental principles of fusion science so that you can evaluate for yourself the future of fusion.

---

[1] Though our current understanding indicates that you will probably never drive a fusion-powered car.

PART 1
MOTIVATION

Chapter 1

The Hydrogen-Powered Civilization

Life needs energy. It always has and it always will. From the simplest bacteria to the most complex civilizations, we need energy to survive.

In this book, we'll take a look at the dominant source of energy in the universe — *fusion*. By smashing hydrogen atoms together to form helium, we can release fusion energy and ensure a sustainable future for ourselves and our planet. However, before we delve into fusion exclusively, the first two chapters will discuss the wide variety of energy sources available to us and fusion's place among them.

1.1 Revolutions in Energy Use

Energy is the basis of modern civilization. The ability to power the objects that enrich our lives is a hallmark of society. Moreover, without the ability to power machines to do useful work, the current world population could not be supported. But how did we get to this point?

Humanity's desire to capture and control external flows of energy predates humanity itself. This is because it wasn't humans, but rather early hominids who first harnessed fire. Though intuitively obvious, fire has three enormous advantages. First, it could be used by the proto-humans to stay warm, thereby reducing the number of calories they needed to consume. Instead of going to the trouble of finding extra food to eat, they could gather nearby vegetation and light it on fire. Second, fire opened an enormous reserve of energy that was previously inaccessible: inedible but flammable material. Finally, it could be used for cooking and light, both tasks that no amount of manual human labor could accomplish. This was the first step in a series of discoveries that unlocked previously inaccessible energy and enabled it to be used in profound ways.

The first agriculturalists used their own bodies to plough fields, dragging the heavy apparatus through the soil. This, however, was a lot of work (according to both the physics and conventional definitions[1]), so humans found animals to do it instead. Oxen and horses, with far more powerful and durable bodies, ploughed our fields and transported people around. Up until the 18th century, animals remained a primary source of power enabling human survival. These beasts of burden were organic machines powered by raw grasses, which human stomachs have difficulty digesting. Though this was a fairly inefficient process, it accessed a new energy source and, since human populations were so low, efficiency wasn't particularly important. Energy abounded, we just needed to find more ways of harnessing it.

In these early times, there were just a few examples of energy manipulation that did not involve living things. Windmills and waterwheels captured the flows of wind and water to process grains and irrigate fields, while the sails of ships enhanced human transportation and exploration. Still, for the most part, we relied on animals and burning plants to enable human survival. This reliance on animals began to change substantially with the invention of the steam engine. The steam engine has a rich and complex history, but here it suffices to say that it became commercially available in the early 18th century. The steam engine was the first device capable of transforming the heat energy from fire into useful mechanical work (see Figure 1.1). While it was initially limited to pumping water out of coal mines, it would eventually power the Industrial Revolution.

---

[1] Work is defined by physicists as the amount of energy used in a process (i.e. the force times the distance over which it is applied).

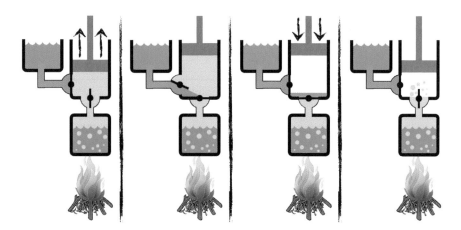

Figure 1.1: A four-frame cartoon of a very early steam engine, which converts heat energy into the motion of a piston. First, water is boiled using the heat, which creates steam that pushes the piston upwards. Next, a valve is opened, letting a small amount of cool water into the chamber to condense the steam. Thus, the piston falls downwards under the weight of gravity. Finally, a valve is opened releasing steam back into the chamber and the cycle is repeated.

Though steam engines were extremely useful, they were only able to do mechanical work.[2] However, in the early 19th century, the British scientist Michael Faraday was to create a new energy paradigm, that of electricity. Faraday discovered *electromagnetic induction* — the fact that an electric current is induced when a conducting material is moved through a magnetic field. Electromagnetic induction was a revolutionary idea that we'll come across again and again in this book. By moving a metal wire past magnets, you can convert mechanical energy into electricity. This works because electricity is carried by the flow of electrons. Since electrons have an electric charge and are very light, the magnetic field strongly affects them and can easily induce them to move. Electromagnetic induction generators became so important because electricity is easy

---

[2]Mechanical work is restricted to moving a macroscopic object, such as a piston. The modern combustion engine in your car mainly does mechanical work, since it serves to rotate the wheels. Similarly, a jet engine does mechanical work by pushing a plane through air.

to transport long distances and can be efficiently converted into other forms of energy. This versatility allowed energy generation to be decoupled from energy consumption, enabling both power plants and appliances to plug-and-play into an electrical grid.

Following electromagnetic induction, humans discovered a number of other ways to generate electricity. For example, the photovoltaic effect directly converts light to electricity, enabling photovoltaic solar panels,[3] and the Seebeck effect directly drives electricity from temperature differences (i.e. thermal energy). Nevertheless, to this day, the vast majority of electricity is still generated by moving a conducting material through a magnetic field.[4]

Many inventors, such as Thomas Edison, realized the usefulness of electrons moving along a wire. Edison's most celebrated product, the incandescent light bulb, uses these electrons in a very clever way.[5] By directing electrons through a resistive carbon filament, they are forced to slow down, thereby transferring the energy to make the filament heat up. Once the wire is sufficiently hot, it radiates visible light! While modern appliances use electrons in far more ingenious and varied ways, the way they are powered remains unchanged — by electromagnetic induction forcing electrons to move along metal wires.

Induction allowed mechanical energy to be converted into electrical energy, enabling the widespread use of electricity. In practice, this is accomplished by rotating a giant, conducting metal disk called a rotor, through the magnetic field generated by magnets. However, how does one get the mechanical energy to rotate the rotor in the first place? For the most part, power plants still use an advanced form of the centuries-old steam engine: the steam turbine.[6] Most

---

[3]When the photons in light strike a material with sufficient energy, they can excite the electrons in the material, driving an electric current.
[4]Strictly speaking, about 1% of all US electricity generation does not rely on electromagnetic induction, namely photovoltaic solar panels.
[5]Edison was not the creator of the light bulb, but he was the first to commercially produce it.
[6]Wind turbines, tidal power, and hydropower are exceptions to this as the movement of the wind/water directly causes the disc to rotate.

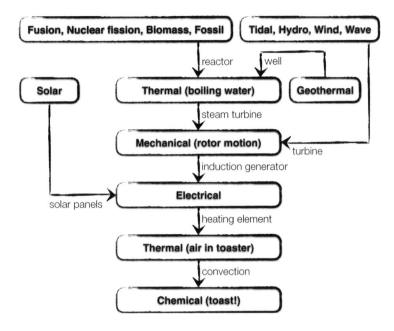

Figure 1.2: How we use the Earth's energy resources to make toast.

energy sources people are familiar with (e.g. coal, oil, gas, nuclear, biomass, and geothermal) are just different methods to create steam. The moving steam transfers energy to the generator rotor, which rotates and induces the electric current. All of this to power your toaster (see Figure 1.2).

1.2 Comparing Options

Selecting what combination of energy sources to use to generate electricity is immensely complex. For a start, the energy reserves of the different sources, as well as their rate of replenishment, vary dramatically. Fossil fuel reserves can only last for a few hundred years and get replenished exceedingly slowly. In contrast, the Earth's solar resources will be continually replenished for the remaining five billion year lifetime of the Sun (at least when it's not cloudy). The amount of fusion fuel is not replenished, but it is so immense that it has the potential to power our civilization for billions of years. Another consideration, the energy density of a source, also varies enormously. To satisfy the current electricity consumption of all Americans using nuclear fission (i.e. the usual nuclear power you hear about) would *only* take hundreds of square miles (the size of a typical city). Solar, on the other hand, would require tens of thousands of square miles (roughly half the size of West Virginia), while wind would need much more (roughly the size of California).[7] However, while these considerations are important, the things we fundamentally care about (things like sustainability, environmental impact, and obtrusiveness) are not so straightforward to quantify.

In order to simplify our decision-making process, we generally let the market decide, assuming that the cost of electricity accurately accounts for the myriad of competing considerations. However, while the market is effective at minimizing costs in the short term, it tends to ignore time-scales longer than the tenure of CEOs. Furthermore, negative externalities (i.e. costs imposed on people who are external to the transaction) abound in energy production. This means that the cheapest source for the individual consumer isn't necessarily the cheapest for society. The most prominent example of this is the emission of carbon dioxide into the atmosphere by burning fossil fuels. When someone buys electricity generated from coal, it imparts a cost, not just on themselves, but on all of society. In emitting greenhouse gases, we are taking out a loan, a loan that must be repaid many times over by future generations. Our descendants will be forced to contend with a changing and less habitable planet. Another externality concerns nuclear fission power, which can aide in the construction of nuclear weapons. As a result of fission generation, governments must use our tax dollars to ensure that nuclear fuel and nuclear waste do not fall into the wrong hands. Moreover, nuclear weapons are just a small example from the much broader *costs of conflict.* Relying on scarce, concentrated energy resources can drive conflicts (e.g. oil wars in the Middle East) that have enormous financial and human costs. And these direct costs may very well

---

[7] Though there is space to do things amongst wind turbines.

be small compared to the effect that war has on international trade and global development.

Regardless, the world is hungry for energy. The Intergovernmental Panel on Climate Change estimates in their baseline scenarios that energy demand will increase by a factor of three by 2100 (and electricity demand will increase from 20% of energy consumption to 50%) [1]. In developed countries, we tend to think of this as a bad thing. To us, it implies that more energy is being wasted or used for superfluous purposes. But across the world, an increase in energy consumption reflects a positive trend — that the global standard of living is rising. Even if developed countries can moderate their energy consumption, the developing world deserves the additional energy needed for health, education, transportation, nutrition, leisure, etc.

In light of all the above, it is clear that we must be deliberate in deciding what energy sources to pursue. It is complex and consequential. We should think deeply and carefully to come to well-informed conclusions. Then, through collective and individual action (e.g. voting, consumer spending, career choice, etc.), we should endeavor to make our desired energy development plan a reality. In the next chapter, we will provide a systematic introduction to the Earth's energy resources. This will serve to guide our thinking on energy policy as well as motivate our investigations into fusion.

Chapter 2

Energy in Numbers and Graphs

Now that we have outlined humanity's history of energy consumption, we'll let the numbers do the heavy lifting as we undertake a systematic review of the dominant energy sources available on the surface of the Earth.[1] But first, what does it mean to "consume" energy? After all, you may have heard that energy is conserved, so where does the energy go? And will energy in the universe ever "run out"?

2.1 Can We Even Consume Energy?

The answer is that energy is never consumed — it is simply converted to other types. However, it can make sense to talk about "consuming" energy (as we will) because not all types of energy are equal. For example, when we heat water in an electric kettle, we are converting electrical energy into heat energy. While we could have used the input electricity for any number of tasks (e.g. powering lights, speakers, clocks, toasters, etc.), the resulting heat isn't so versatile.[2] Operating a kettle does not destroy any energy, but it does cause some energy to become less useful.

The principles underlying the usefulness of energy are quite technical and stem from the *entropy* of different energy types.

---

[1] For a 400 page version of this, we highly recommend *Sustainable Energy Without the Hot Air* by David MacKay.
[2] This is why we don't charge our cell phones with kettle spout → USB adapters. Or maybe it's just because Apple has yet to enter the tea market...

Entropy is a rigorous way of quantifying the amount of order in a system — the more *disordered* the system, the *higher* the entropy. Physics tells us that the entropy of an isolated system will increase with time. In other words, things tend to chaos. Types of energy with low entropy (like electricity) are more useful because we can take advantage of their order to do things that we want. The directed march of electrons down a wire can be manipulated much more purposefully than the haphazard and random motion of a hot gas. Again, this isn't to say that heat energy is always completely useless, just that it is less useful per unit of energy. This is seen most clearly from air conditioners used to cool homes. Air conditioners can use one unit of electrical energy to move four units of heat energy from inside your house to outside. This contrasts with steam-driven induction generators, which generally require around three units of heat energy to generate just one unit of electrical energy.

Thus, when we talk of "consuming energy," what we really mean is "producing entropy." The forms of energy that can power our civilization (i.e. our "energy sources") are able to do so because we can convert them into higher entropy forms. This process of entropy generation is inescapable and provides a fundamental limitation. We must increase entropy to accomplish the things we want, but entropy can only be increased so much. Eventually, we will run out of low-entropy fuel and will be left with only useless high-entropy energy. Luckily, the universe provides sufficient low-entropy energy sources to last for ages, which we will discuss now. All we have to do is harness them.[3]

2.2 A Brief History of Energy

About 10 minutes after the Big Bang, the universe rapidly cooled, allowing atoms to form out of a soup of free protons, neutrons, and

---

[3]The amount of useful energy available in the universe appears limited. It appears that, no matter what we do, the universe will eventually reach "heat death" and nothing more complex than low-energy photons will exist. This is not predicted to happen for a very very very long time, so don't worry. Also, it might be wrong.

electrons. Because of this deconstructed beginning, the universe could only make the very simplest of atoms. Looking closely you would have seen that roughly 90% of the atoms made by the Big Bang were hydrogen, the smallest atom in the Periodic Table of the Elements, while the rest were helium, the second smallest. Starting from this initial condition, a supermassive cloud of hydrogen and helium, we will trace the flows of energy shown in Figure 2.1 and arrive at the important features of the Earth's current energy budget.

A cloud composed of light atoms has two dominant sources of usable energy: fusion and gravity. As we will explain shortly, fusion is the process of combining light elements to form heavier ones. This process releases energy, a lot of energy. You can take two atoms of hydrogen and combine them to produce one atom of helium, which will be moving much faster than the hydrogen was. The energy released by the reaction is contained in the kinetic energy of the helium. The other energy source in our cloud is gravity. If you take two atoms and place them in the frictionless vacuum of

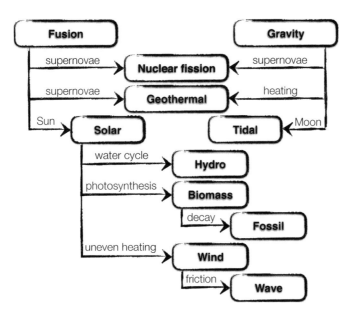

Figure 2.1: The energy hierarchy: a summary of the flows of energy through the universe.

space, they will pull on each other through gravity. As they move closer to each other, they gradually speed up, exactly the way a dropped cell phone accelerates towards the cold, hard surface of Earth. In this way, gravity causes a cloud of gas to collapse in on itself, while simultaneously increasing the average speed of the atoms in the cloud. In everyday life, we call "increasing the average speed of atoms" by another name: heating.

There is one important caveat concerning fusion: atoms are not easy to combine. In fact, successfully fusing atoms requires specific and extreme circumstances. Normally, if you have a cloud of atoms, they will simply bounce off each other and never fuse. Only when the cloud is very hot, roughly $10,000,000°C$,[4] is there a decent chance that fusion will occur. Billions of years ago, the universe was not this hot, so fusion energy was entirely inaccessible. Hence, gravity ruled.

Gravity is the spark that enabled *everything*. Over billions of years gravity expended prodigious amounts of energy into collecting and compressing the atoms from the formerly smooth and diffuse universe. As these clouds were formed and collapsed in on themselves, gravitational energy was transferred to thermal energy in the gas. The effect of the astronomically large number of particles, all falling towards each other, was to produce the temperatures necessary for fusion. Thus, stars were born. Stars convert hydrogen into helium, generating the vast quantities of energy that they radiate into space in the form of sunlight.

After millions or billions of years, stars eventually burn through all their fusion fuel and undergo further gravitational collapse. If the star is large enough, this will result in an incredibly energetic explosion called a supernova. Supernovae are so energetic that they are able to form heavy elements. Since our Sun isn't the first star in the universe, some of these heavy elements were able to find their way into the nebula that formed our solar system. This small amount of dirt (less than 0.1% of atoms) seems inconsequential, but we will see

---

[4]Or °F, or Kelvin, or Rankine. With temperatures this extreme the units don't end up mattering much.

that it has the potential to fundamentally alter humanity's energy future.

TECH BOX: Heavy element synthesis

Strictly speaking, there are two dominant methods that create "heavy" elements (i.e. elements heavier than iron). The r-process, or rapid-process, only occurs when a star goes supernova. This is fairly uncommon because supernovae require the death of a very massive star (typically at least ten times more massive than our Sun). However, the r-process is so effective that it is still responsible for about half of all heavy elements. The other half is produced by the s-process, or slow-process, which occurs in intermediate-sized stars during their red giant phase. Both processes work by bombarding medium-sized elements like iron with a large number of neutrons. The difference is that the r-process is so rapid that the elements don't have time to undergo radioactive decay between capturing neutrons.

In our own Solar System, gravity formed over 99% of all matter into a single enormous celestial sphere: our Sun. Because of its immense mass, it will always dominate the energy budget of the Solar System. Additionally, gravity created many smaller objects: the planets and their moons. Though none contained enough atoms to achieve the high temperatures needed for fusion, we still like them.

This story is sufficient to trace the dominant flow of energy through our Solar System. The potential energy of our proto-Solar System Nebula was dominated by fusion and gravity. Initially gravity reigned supreme, compressing and heating our nebula to form the Sun and planets. This process converted much of the gravitational energy into thermal energy, enabling the Sun to achieve the necessary temperatures for fusion. Since then, and for the next 5 billion years, fusion energy will dominate our Solar System.

Earth, however, is a unique place and requires special investigation. Like most places in the Solar System, we receive a lot of solar energy from the Sun's fusion. We can also still access some of the original gravitational energy of our nebula. This is because gravity hasn't quite finished forming the Earth–Moon system. By siphoning off energy from the tides using tidal power plants, we can capture some of it before it is lost to thermal energy of the oceans. However, what makes the Earth really different from the Solar System as a whole is that it has trouble holding on to hydrogen. Planets that are close to the Sun (i.e. Mercury, Venus, Earth, and Mars) are sufficiently warm that hydrogen can sometimes escape from solid materials (e.g. ice, ammonia, and methane). Since an individual hydrogen atom is so light, once it is free, any incidental bump overwhelms the force of gravity and sends it flying off into Outer Space. This has the effect of dramatically increasing the relative abundance of heavy elements, making Earth's energy budget special. Fortunately for us, some hydrogen still remains, safely locked away in liquid water (i.e. H_2O). So while we still have tremendous stores of fusion energy, they aren't nearly as dominant as on, say, Jupiter.

Though negligible in the Solar System as a whole, heavy elements represent a substantial source of energy here on Earth. These elements stored fusion and gravitational energy from distant supernova that occurred billions of years ago. They then found their way into the Earth where they undergo radioactive decay and help to keep the core of our planet hot.[5] We can access this energy (as well as residual gravitational energy from the Earth's formation) by digging deep wells as part of geothermal power plants. Alternatively, we can directly mine the heavy elements (most notably uranium) and use them to fuel nuclear fission power plants.

The above are the "fundamental" energy sources here on Earth: fusion, solar, tidal, nuclear fission, and geothermal.

Hold on, what about wind energy? Well, there are a number of energy sources like wind that we will separate from these fundamental sources. Most of these "by-product" sources are directly driven by

---

[5]Specifically the decay of thorium, uranium, and potassium produces most of the Earth's underground heating. Since potassium is smaller than iron, it does not require supernovae to be created.

solar energy (see Figure 2.1). Distinguishing them is useful because it informs our intuition. We know that a by-product will necessarily contain less energy than the more fundamental energy source that is continually powering it. For example, we use wind turbines to capture the energy in the wind. However, the dominant cause of wind is the difference in the solar energy received by the equator compared to the poles. Wind, in turn, can push on the surface of the ocean and transfer the energy needed to create waves.

Another by-product energy source, hydropower, works by using dams to extract energy from water running downhill. However, the entire water cycle is ultimately driven by evaporation, which is caused by sunlight heating the surface of the oceans. Lastly, fossil fuels and biomass both depend on plants, which use photosynthesis to capture sunlight. Biomass involves growing plants to immediately burn for energy, while fossil fuels are the remains of ancient plants, which captured sunlight millions of years ago.

Figure 2.1 summarizes the flows of energy that determined the major energy sources here on Earth.[6] These sources of energy can serve as inputs into Figure 1.2 in order to put electricity on the grid and toast on your plate. Figure 2.1 also presents a clear hierarchy, which indicates the relative size of various energy sources. More fundamental sources in the hierarchy tend to dominate on average in the Solar System. However, here on Earth things are a bit special, so we must investigate the specifics. How much of the Sun's energy does the Earth intercept? Is there still much gravitational energy left after the formation of the planets? What is the abundance of hydrogen and uranium on the Earth's surface?

2.3 Our Energy Resources

In this section, we will review each energy source from Figure 2.1. We will run the risk of over-simplifying by restricting our analysis to

---

[6]This is a simplified picture that neglects many smaller stores and flows of energy. For example, geothermal energy, through plate tectonics, forms mountains and feeds energy back into gravity.

back-of-the-envelope calculations concerning only the most essential features. This is done to establish the general size and shape of the global energy problem and determine fusion's position within it. *The size of each energy source will be compared to humanity's total rate of energy consumption: 18 terawatts* (see the "Units of energy and power" Basics Box). This is the entire primary energy supply for the world, which includes everything: electricity generation, transportation, heating, and all thermodynamic/transmission losses. This value is roughly three times greater than the current *electricity* consumption because heating and transportation currently rely on the direct combustion of fossil fuels.[7] Humanity's total energy consumption will likely change significantly in the future. Increasing the efficiency of production, transmission, and consumption tends to lower our energy requirements. However, historically, any gains in efficiency have always been dominated by increases in our use of energy. Considering that per capita energy use varies by a factor of 10 between developed and underdeveloped countries, global energy use will likely continue to increase, even if developed nations can manage to curb their consumption.

BASICS BOX: Units of energy and power

A common source of confusion about energy generation is the distinction between energy and power, as well as the various units for each. Energy is what we're fundamentally after. It's the substantial quantity. A pile of charcoal has a certain amount of energy. This can be measured in Joules (J), which is a pretty small unit. For example, a kilogram of charcoal contains 30 million Joules (abbreviated as 30 megajoules or 30 MJ). Power, on the other hand, is the rate of energy consumption. So if you burn the kilogram of charcoal quickly, it will sustain a higher level of power, but for a

(Continued)

---

[7]It seems likely that transportation (and to a lesser extent heating) will soon rely much more on electricity.

(*Continued*)

> shorter time. Power can be measured in units of Watts (W), which is exactly equivalent to 1 J/s. So burning a kilogram of charcoal over the course of 3,000 seconds would continually produce 10,000 Watts (abbreviated as 10 kilowatts or 10 kW). A Watt is a pretty good size for our power consumption. A lit incandescent light bulb continually consumes roughly 100 Watts or 0.1 kW. A large power plant produces about 1 billion Watts of electricity (abbreviated as 1 gigawatt or 1 GW). The human species consumes about 18 trillion Watts on average (abbreviated as 18 terawatts or 18 TW). Because the Joule is so small, it is common to use the Watt to define a new unit of *energy*: the kilowatt-hour (abbreviated kW-h). This is exactly equivalent to the amount of energy consumed by operating a 1 kW appliance for 1 hour.

In comparing different energy sources, we must be careful because some have fixed stores of energy that are never replenished (e.g. fusion and fission), while others are continually replenished (e.g. solar, wind, biomass, hydroelectric, and wave). The distinction between these categories is at the heart of what separates "renewable" energy sources from "non-renewable" ones (which we will briefly discuss at the end of this chapter). Here, we will treat them somewhat differently. For non-renewable sources, we will calculate how long they could sustain our current level of power consumption (i.e. 18 TW). For renewable sources, we will calculate how their rate of replenishment compares to our current level of power consumption.

2.3.1 *Fusion*

Fusion is the process whereby the nuclei of light elements such as hydrogen and helium are combined together to produce energy. We will explain this process in detail in Chapter 3, but, to estimate Earth's fusion energy resources, it is sufficient to know just two of the most attractive fusion reactions.

The first is the reaction that will almost certainly power our early fusion power plants. This reaction is the easiest and combines two *isotopes* of hydrogen called deuterium and tritium. Different isotopes of the same element have the same number of protons in their nucleus, but a different number of neutrons. So, normal hydrogen has a nucleus composed of a single proton, deuterium has a nucleus with one proton and one neutron, and tritium has one proton and two neutrons. The deuterium–tritium (D–T) reaction produces helium-4 (a nuclei with two protons and two neutrons) and one free neutron. This reaction is the most attractive because it has the highest probability of the two nuclei fusing (as opposed to scattering off one another) and this probability occurs at a *relatively* low temperature.

While deuterium occurs naturally and can be found in every glass of water, tritium is harder to come by. It is radioactive and, if you try to store it, half of it will decay into helium every 12 years — a characteristic time known as the isotope's *half-life*. So, if you started with 100 grams of tritium, after 12 years you would only have 50 grams and after another 12 years you would be down to just 25 grams. Hence, nature doesn't have much tritium, so we must make it ourselves. This is not ideal, but it can be done using what are called tritium breeding reactions. Typically, this works by surrounding the fusion reactor with lithium and using the free neutron produced by D–T fusion to transform it into a tritium atom and a helium atom. So, in summary, even though deuterium and tritium are being fused in the reactor, the overall inputs to a D–T fusion power plant are deuterium and lithium.

The second reaction is harder, but can presumably be achieved after humanity becomes an expert in fusion. In this reaction, known as deuterium-deuterium (D–D) fusion, two deuterium atoms are fused. D–D fusion requires higher temperatures and, even so, the likelihood of the atoms sticking (as opposed to scattering) is much smaller than D–T. As a result, D–D fusion power plants must be bigger and better to achieve the same performance. The primary selling-point is that breeding tritium from lithium is not required, so the only input to the power plant is deuterium.

One of the many reasons why fusion is attractive is its insanely high specific energy. Specific energy, sometimes called energy density, is the amount of energy a fuel contains *per unit mass*. Since fusion produces energy by modifying the nuclei of atoms (i.e. nuclear energy) as opposed to modifying electron orbits (i.e. chemical energy), we expect the specific energy of fusion fuels to be a million times larger than fossil fuels.[8] This is because the force that keeps the nucleus together is a million times stronger than the force that keeps electrons in orbit. Moreover, even when compared to nuclear fission, fusion has a significantly higher specific energy (by about a factor of four). Because of this, *much of our intuition about fuel use is no longer accurate when considering fusion.* Instead of using a kilogram of charcoal over the course of an evening barbecue, you use a kilogram of fusion fuel to power the complete lifelong energy needs of you and your closest 100 friends. Put another way, the lithium currently in your cell phone battery and the deuterium currently in your body (a few tablespoons) would suffice to satisfy your energy needs for over 10 years. Clearly, a small amount of fuel goes a long way! Consequently, issues like the cost and environmental impact of getting the fuel tend to be dramatically smaller than for fossil fuels. These concerns, which are often the first thing our mind jumps to, are not particularly important for fusion.

To estimate our fusion fuel supply, let's first look at lithium, which turns out to be the limiting resource for D–T fusion. It is currently mined in the Earth's crust, but can also be extracted from seawater (albeit at a higher price). In 2010, the US Geological Survey estimated the world's potentially extractable lithium deposits to be roughly 30 million tons [2]. This is enough to sustain humanity's current rate of energy consumption (i.e. 18 TW) for 10,000 years. Expanding our view to the oceans, we find that every liter of seawater

---

[8]The higher specific energy of fusion compared with other energy sources does not imply that the specific energy of a fusion *power plant* will necessarily be higher than for other sources. It just means that we need far less fuel to generate a unit of energy. This is very useful because gathering and transporting smaller amounts of fuel tends to be cheaper. Moreover, any given fuel reserve will last longer.

contains 180 micrograms of lithium. While this doesn't seem like much, the oceans are so vast that they contain 200 billion tons of lithium. This is sufficient to fuel D–T reactors providing our current energy consumption for a 100 million years.[9] However, this all assumes that other industries (most notably the lithium-ion battery industry) don't take all the lithium first. Fortunately, because fusion is a nuclear process, while battery storage is a chemical one, it looks like fusion would be able to pay much more than the battery industry.[10]

Moreover, we only need enough lithium to enable a transition to D–D fusion. Deuterium was directly formed in the Big Bang and is the third most abundant isotope in the universe. Here on Earth, deuterium makes up 1 out of every 6,400 hydrogen atoms. It is fairly easy to filter out of sea water, making it cheap and accessible to all. Even now, with just $20 and an internet connection you can buy enough deuterium to cover your lifetime energy needs. The oceans alone contain more than enough deuterium to provide over 10 billion years of energy at humanity's current level of consumption. Furthermore, it appears practical to consume nearly all of this because deuterium and hydrogen are more or less chemically indistinguishable. As far as ocean ecosystems can tell, we are simply removing 1 out of every 6,400 hydrogen atoms: an infinitesimal perturbation. Since the ocean mixes every few thousand years [3], we can be sure that the deuterium currently near the ocean floor will make its way to the surface before we need to use it.

---

[9]Here you might object and point out that extracting lithium from the oceans could very well be much more expensive. However, the response is "nuclear fuel." Fusion fuel costs are expected to be so cheap that we don't care too much if seawater extraction is more expensive.

[10]Two lithium batteries from the Tesla Powerwall are reported to contain about 25,000 grams of lithium and retail for $10,000. In a fusion power plant, 25,000 grams of lithium would enable you to generate 2 TW-h of energy, which you can sell for about $200,000,000. While this isn't terribly rigorous, these prices indicate that fusion power plants should be able to outbid battery companies, even if batteries are recycled many times.

2.3.2 Nuclear fission

Fission and fusion are different. Most noticeably, fission currently generates over 10% of the world's electricity, while fusion generates 0%. This is primarily because the following two statements are true:

<u>Fusion</u> energy generation is <u>difficult</u> to start, but <u>easy</u> to stop.
<u>Fission</u> energy generation is <u>easy</u> to start, but <u>difficult</u> to stop.

Due to the relative ease of inducing fission reactions, fission power plants required less research and development, quickly entering the electricity market. However, the difficulty of turning off a fission plant is a compelling reason to prefer fusion. To understand why fission is hard to stop, we must first start with the basics. As the name suggests, nuclear fission involves breaking one heavy atom into two lighter ones. The most practical way to do this is to hit certain barely stable nuclei, most notably that of uranium-235 (composed of 92 protons and 143 neutrons), with a neutron. The extra neutron and the energy it carries makes the nucleus unstable, causing it to break apart. Usually, two *fission products* and a few free neutrons are released, all with enormous kinetic energy. To produce sustained power, the reactor geometry is set up so that, on average, exactly one of these free neutrons strikes another uranium-235 nucleus and causes it to fission. This is called a nuclear chain reaction. The kinetic energy from all of these fissions can be used to heat water, run a steam turbine, and generate electricity.

However, while the kinetic energy carried by the fission products/neutrons is released instantaneously, it isn't all of the energy. When a large atom splits, it can divide into any of hundreds of nuclei, but most of them are radioactive (i.e. they are unstable). Depending on the exact identity of the fission product, it could wait for seconds or for millions of years before it decays and releases some amount of additional energy. This delayed energy is intrinsic to the physics of fission power and prevents a power plant from being turned off instantaneously. An hour after all fission reactions are stopped, this process of radioactive decay is still producing over 1% of the reactor's operating power. 1% may not sound like much, but when the flow of coolant stopped in Three Mile Island and Fukushima, it was enough to melt the reactor core.

Eventually, after a few days, the power released by radioactive decay becomes negligible *from a power plant cooling perspective.* However, there are still long-lived fission products that continue to decay, which can release particles and/or electromagnetic radiation with high energy. For example, a technetium-99 atom waits 300,000 years on average before it decays, releasing two energetic electrons. If ingested this poses a health risk to humans, which has motivated the US government to look for ways to safely store spent fuel for a million years.

Additionally, it is *critical* that, on average, no more than one neutron from each fission go on to cause another fission. If the average is 1.01, then the chain reaction can spiral out of control much too quickly to stop, causing a steam explosion. Obviously, this is a very serious concern as it can breach the containment structures and exhaust radioactive material to the environment. This must be addressed by ensuring *negative feedback* in all circumstances. Negative feedback means that an increase in power makes it less likely for neutrons to find fuel nuclei. This was not the case for the infamous Chernobyl power plant, which had a design adapted from reactors optimized to build nuclear weapons.

Lastly, before we estimate the world's fission fuel resources, we must also mention that nuclear fission power plants (and the spent nuclear fuel they produce) can facilitate the construction of nuclear weapons. We will delay a more in-depth discussion of nuclear weapons proliferation and how it relates to both fission and fusion until Chapter 10. Here, it will suffice to say that this is a cause for concern and is, in our opinions, the most compelling argument against fission power.

The usual fuel for nuclear fission power plants is the heaviest element that is present in nature: uranium. Current fission power plants overwhelmingly use the isotope uranium-235 because it can be fissioned directly by a single slow-moving neutron. However, this isotope makes up less than 1% of natural uranium. The other 99% is uranium-238, which is more difficult to work with because it generally requires two neutrons to fission. The first neutron is used to induce a breeding reaction, which converts the uranium-238 atom into plutonium-239, and the second to fission the plutonium. Currently, there is only one commercial breeder reactor in the world, though there have been a number of experimental devices. This motivates two fuel estimates. First, how much energy can be generated using our current strategy of burning uranium-235? Second, how much can be generated using advanced breeder reactor technology?

There are roughly 10 million metric tons of *identified and economical* uranium deposits buried across the world. Since uranium has never been a particularly scarce resource, much more than this almost certainly exists (a factor of 1,000 more has even been suggested). Let us assume that we can eventually find 100 million tonnes of raw uranium. How much energy does that give us? Well, uranium is a nuclear fuel, which means it is very energy dense. One ton contains roughly the weekly electricity consumption of Germany.[11] That means that we would have enough energy to sustain the world's current needs for about 10,000 years. Not too bad! However, this requires advanced breeder technology. With current technology, the energy held by uranium-238 is largely inaccessible. Using only uranium-235 (which is 1% of natural uranium), the world's *identified and economical* deposits will last for just 100 years.

Luckily, there is an even bigger source of uranium: the oceans. Although every kilogram of ocean water only contains 3×10^{-9} kilograms of uranium, there are over 10^{18} kilograms of ocean water (for a refresher on scientific notation see the Basics Box). This represents a vast source of energy.[12] Given advanced breeder technology, it would be capable of sustaining our current

---

[11] This can be found by noting the fission of a 235 AMU uranium atom releases roughly 200 megaelectron-volts, then converting to more macroscopic units.

[12] A much-discussed alternative fission fuel is thorium (90 protons and 142 neutrons). It is three times as abundant as uranium in the Earth's crust and has better non-proliferation characteristics than uranium. However, similar to uranium-238, it must undergo a breeding reaction (into uranium-233) before being fissioned. While there is more thorium than *terrestrial* uranium, the oceans contain the dominant supply of potentially accessible fission fuel. Since thorium

level of consumption for almost 1,000,000 years. Given its minuscule concentration, you might doubt our ability to economically extract uranium from ocean water. Fortunately, we are dealing with a nuclear fuel. Because of its incredible energy density, the current cost of uranium in existing fission power plants is only about 10% of the overall cost of electricity. Hence, even if switching from mining to seawater extraction made the price of uranium triple, the overall cost of electricity would increase by just 20%. Moreover, the fuel costs for breeder reactors are expected to be almost a hundred times smaller because they burn all the uranium (not just uranium-235). Though techniques to extract uranium from seawater are still in the research and development stage, the initial cost estimates appear reasonable (at least for breeder reactors) [4].

In conclusion, our current use of uranium, once-through burning of terrestrial uranium-235, is able to satisfy 100% of our energy needs for a century or so. If breeder power plants and uranium seawater extraction can be successfully implemented on a large scale (including managing nuclear weapon proliferation concerns), our energy problems can be considered solved.

BASICS BOX: Scientific notation

Throughout this book, we will use scientific notation to write very large and small numbers. The number in the exponent tells you how many zeros are in the number. If it's positive, then you add that many zeros after the 1 and, if it's negative, you add that many zeros before the 1 (including the one that precedes the decimal point). So $10^2 = 100$ has two zeros, $10^4 = 10,000$ has four zeros, and $10^{-3} = 0.001$ has three. If there is a number out front, then you multiply it after you've put the appropriate number of zeros. Hence, $4 \times 10^3 = 4,000$ and $1.5 \times 10^{-1} = 0.15$.

---

is not water-soluble, most of it stays locked away in rocks at too low of a concentration to be economically practical.

2.3.3 Geothermal

Geothermal energy is driven by two mechanisms, in roughly equal proportions. First, in forming the Earth, gravity caused a large amount of matter to collapse in on itself. This process converted a tremendous amount of gravitational energy into thermal energy of the newly formed Earth. Second, the Earth's crust and mantle contains radioactive heavy elements formed in supernovae during the deaths of stars long ago. These heavy atoms (mostly thorium and uranium) gradually decay into stable elements, releasing large amounts of nuclear energy by fission and radioactive decay.[13]

The geothermal heat from these two sources gradually conducts outwards to the Earth's surface and is radiated into space (along with all of the energy received from the Sun). However, we can intercept it along the way and use it to generate electricity. To accomplish this, we dig wells several kilometers deep and use the high temperatures to power a steam turbine. Unfortunately, the Earth is a poor thermal conductor, so the heat leaks up to our wells very slowly. We can calculate exactly how slowly by observing that, 40 kilometers below ground, the temperature is roughly 600°C. By multiplying this temperature gradient (i.e. 0.015°C/m) by the thermal conductivity of rock (i.e. 2 W/m/°C), we see that 0.03 Watts of geothermal energy is continually passing through each square meter of the Earth's surface. When totaled over the whole planet, we find that it barely exceeds humanity's current total power consumption. Thus, although the Earth contains enough geothermal energy to satisfy our current consumption for 10 billion years, we can't take advantage of it. The energy conducts to the surface so slowly that we can't really access it. If we try to draw energy more quickly it becomes unsustainable as it will cool down the rocks surrounding our wells and reduce the geothermal energy available in the future. Unless we manually increase the bulk thermal conductivity of the Earth to allow the heat to escape faster (e.g. millions of wells that span a sizable fraction of the Earth's radius), this represents a fundamental limit on geothermal energy.

---

[13]This radioactive decay is the same process that prevents fission power plants from turning off quickly.

While we certainly should harness all economically competitive geothermal energy, these numbers demonstrate that we should never expect it to provide a majority of our energy. Geothermal is great if you happen to be sitting on a geothermal hotspot (e.g. Iceland), but not so much for the rest of us. It can barely satisfy our current energy demands, even assuming that it can be completely captured. For the most part, geothermal energy is diffuse and homogeneous, so the idea that even 10% of it can be extracted stretches the imagination. After all, 70% of geothermal energy exits the Earth through oceans, which make drilling problematic.

2.3.4 Solar

The critical reaction that enables energy production in the Sun is the fusion of two normal hydrogen atoms to form deuterium. This is known as proton–proton fusion because the nucleus of normal hydrogen is a proton. Despite the vast amount of normal hydrogen, we do not pursue this reaction here on Earth for the simple reason that it appears impossibly difficult. In the easiest fusion reaction, deuterium and tritium bounce off one another 100 times on average before successfully fusing. This seems like a lot. Now consider that proton–proton fusion will see roughly 10^{30} such scattering collisions before fusion! The Sun can overcome this infinitesimal probability because of its astronomical size and high density compared to terrestrial fusion devices.

TECH BOX: Stellar fusion

The probability of proton–proton fusion is so small because it relies on the weak nuclear force. When two protons fuse they form a diproton (i.e. the isotope of helium with zero neutrons). This nucleus is unstable and usually decays back to two separate protons immediately. However, the weak nuclear force can cause one of the protons in the diproton to spontaneously transform into a neutron via positron emission (i.e. emitting a

(Continued)

> (*Continued*)
>
> particle with the weight of an electron, but a positive electric charge). This produces deuterium, which is stable, so fusion has been achieved. Nevertheless, as indicated by the name *weak* nuclear force, converting a proton into a neutron during the fleeting existence of a diproton is extremely unlikely.

Like all types of fusion, the energy produced by proton–proton fusion is released in the kinetic energy of the products. This, in turn, is converted to heat as the products collide with the other particles in Sun. This heat eventually makes its way to the surface and is radiated out into space as sunlight. While only 1 part in 10^{10} of the Sun's light hits Earth, we still receive an enormous amount, more than enough to power life as we know it.

The Earth's surface receives 10^{17} W of solar energy from the Sun, or about 5,000 times humanity's total power consumption [5]. This averages to roughly 200 W/m^2, which makes sense since a few 100 Watt lightbulbs can illuminate a square meter about as well as the Sun. While this is large, we can't really expect to take all of this energy. After all, solar is the primary drive of nearly all activities on our planet. If we want the Earth to retain its essential features (night/day, seasons, sunburn, plants/animals, etc.), we cannot hide the planet under the permanent shade of a single global photovoltaic panel. While no one is advocating this, it forces us to reduce our expectations of solar somewhat. The environmental impact of the widespread adoption of solar power doubtlessly depends on the details of how it is implemented. Nevertheless, it seems clear that the vast majority of solar energy must be left untouched so it can power the planet, rather than human civilization. Still, even after allocating just 2% of the Earth's solar resources to humans, we are left with 100 times our current energy consumption. More importantly, we get to choose which 2% to use for energy generation. This means we can select only the most economically optimal locations (i.e. sunny places) that also have minimal environmental impact. This is a big selling point for solar. Because there is so much of it, we can afford to be picky about which sites we exploit. Clearly, there is enough solar energy to be a sustainable solution.

Like all energy sources, solar has some drawbacks. First, though recently prices have dropped significantly, rooftop photovoltaics and solar thermal are still some of the most expensive methods of electricity generation (though large, centralized photovoltaic installations are more competitive) [6]. Second, solar energy is intermittent. Nights happen every day, winters happen every year, and sometimes it's cloudy. This intermittency, most notably the decrease during winter, is a serious concern. There may be practical solutions, but they will likely be expensive. We will delay an in-depth discussion of intermittency until Section 2.4. Lastly, photovoltaics are currently one of the least "green" alternatives to fossil fuels. Life cycle analysis studies [7] indicate that photovoltaics release nearly three times as much carbon as nuclear fission and four times as much as wind (though it is still about ten times better than natural gas). Fortunately, this largely results from the energy required for their manufacture. Hence, the carbon footprint of photovoltaics can be expected to shrink as electricity generation is decarbonized.

2.3.5 *Wind*

Because the Earth is spherical, the equator receives much more solar energy than the North and South poles. This creates a temperature difference that is the dominant force driving the wind. Interestingly, this process has close parallels to fusion, so understanding the wind will help us prepare for discussions of turbulence in fusion devices. Here the question we want to answer is "How much wind does this solar-induced temperature difference create?" In other words, if the Earth receives 100 units of solar energy, how many units of wind energy arise?

To understand the wind, we can use the knowledge of thermodynamics that we gained in discussing the steam engine. This may seem surprising, but for both we need to know how efficiently chaotic thermal energy can be converted into directed kinetic energy. While an optimized steam engine can achieve real-world efficiencies above

40%, we will see that the Earth's atmosphere is not so well designed. This is because the most important thing for high efficiency is a large temperature difference (see the following Tech Box). So while the steam produced in power plants is typically 400°C hotter than the room temperature water used as a coolant, the Earth's equator is only a measly 60°C hotter than the poles. But wait, there's more!

TECH BOX: Thermodynamic efficiency

The Carnot efficiency is a useful metric for estimating how much directed kinetic energy can be produced from a certain amount of thermal energy. This efficiency can never be achieved in practice, but instead is a theoretical upper limit. It's a best case scenario. The Carnot limit on efficiency is given by

$$\eta \leq 1 - \frac{T_C}{T_H}, \qquad (2.1)$$

where T_H is the temperature of the hot reservoir (in units of absolute temperature, like Kelvin), T_C is the temperature of the cold reservoir (also in absolute units), and η is the thermodynamic efficiency we want to calculate. For the case of a coal power plant steam cycle, T_H would be the temperature of the steam produced by the reactor, T_C would be the temperature of the coolant used to condense the steam (so it can be reinjected into the reactor), and η would be the kinetic energy of the spinning turbine divided by the thermal energy produced from burning the coal. Large optimized power plants often achieve two-thirds of this ideal efficiency.

Because the Earth spins, air near the equator has to move east at 1,700 kilometers per hour to be measured as "stationary" with respect to the ground, while "stationary" air at the poles really is stationary (relative to the center of the Earth). Hence, even though we don't notice it standing on Earth, there is a large velocity difference between the flow of air at the equator and the air at the poles. Because of this, when you are flying in a plane traveling north in the northern hemisphere, it actually has to point slightly west just to stay on the same line of longitude. This phenomenon, known as the Coriolis force, prevents hot air at the equator from directly traveling to the poles. As soon as a chunk of air from the equator starts to move north, its now-excessive eastward velocity makes it veer sideways (see Figure 2.2). If the Earth were not spinning, a single enormous convective eddy might span the entire distance from equator to pole and much more efficiently equalize the temperature difference created by the Sun. Instead, the Earth's rotation shears the one large eddy into three smaller ones (in both the northern and southern hemispheres) as shown in Figure 2.2.[14] This reduces the

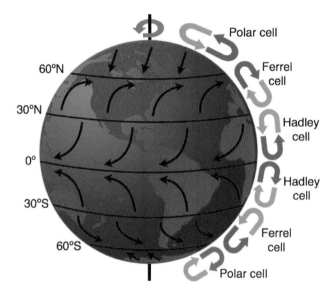

Figure 2.2: A simplified cartoon of the six large-scale convective eddies that dominate the global wind patterns, along with the surface winds they create.

---

[14]These three eddies are now known as the Hadley, Ferrel, and Polar cells. Back when maritime navigation was at the cutting-edge of science and the ocean was the frontier of exploration, these weather patterns were not well understood, but played a much more prominent role in human culture. Terms like the trade winds

amount of wind energy because it is now being generated by 20°C temperature differences, not a single 60°C difference.

Using a 20°C temperature difference, Equation (2.1) tells us that the highest possible efficiency is roughly 5%. In other words, at most 5% of the solar energy that heats the atmosphere is converted to wind energy. More detailed models created by climate scientists estimate that about 1% of solar energy is converted into wind [8].

At 1% of solar energy, the Earth's winds still contain 50 times our current energy consumption. However, the 1% efficiency also means that wind energy is very diffuse. On average, a given area of the Earth's surface can provide 100 times more solar energy than wind. Moreover, solar energy can be entirely intercepted by a flat horizontal plate at the Earth's surface, while wind energy is dispersed over the entire 10 kilometers extent of the troposphere. Fortunately, the difficulty in collection is more than compensated for by the simplicity of wind energy technology. Humans have been harnessing the power of the wind for thousands of years, so, while gains in efficiency are ongoing, it is fair to say that we are wind experts.

Like solar, wind provides a number of functions that are essential to the planet as we know it (e.g. weather/climate, awesome birds of prey, dispersing pollen, etc.). However, because it's so diffuse, wind power will become economically infeasible long before we capture enough of it to bother Mother Nature. Yet, it does seem reasonable to think we could extract the most convenient percent or two of the Earth's wind energy. This estimate for our wind generation capacity is almost exactly equal to humanity's current total energy demands, so it is important to improve its accuracy as much as possible.

Detailed studies estimate that roughly four times our current energy consumption can be generated by reasonably tall wind turbines (including off-shore resources) [9] and about a quarter of this

---

(the surface air flows of the Hadley cells) the doldrums (the stagnate region where the two Hadley cells meet at the equator), and the horse latitudes (the stagnate region where the Hadley cells meet the mid-latitude cells) have meanings that are now all but forgotten. Still they possess the charm of an era where the world held much more mystery.

currently appears economic (i.e. has an estimated cost of electricity <$0.09/kWh). However, these studies don't include collective effects, which look like they may become important. When you start to build truly enormous wind farms (i.e. thousands of square kilometers) you can extract a significant amount of the *local* wind energy. These effects are local to the wind farm, but have the potential to significantly reduce total generation, maybe by a factor of four [9]. This indicates that, to get the amount of energy humanity requires, we have to push wind power to its limits. We have to build expansive wind farms everywhere remotely appropriate and with sizes such that they begin to affect the *local* wind speeds.

In order to generate this amount of wind power, turbines would have to become fairly ubiquitous. A typical large wind turbine is over 100 m tall, can be plainly seen from 10 kilometers away, and has an average electrical output of 1 MW. If we want to provide 1 TW of power, less than a third of the United States' current energy consumption, we would need 1 million turbines. To put this in context, it is three times larger than the number of traffic signals in the country. If they were spaced evenly across the US you would be able to see more than 30 turbines from any place with clear visibility. If they were concentrated into dense wind farms,[15] they would span roughly 10% of the country's land area.

Wind generation has seen a dramatic expansion over the last 10 years because it is already economically competitive with other generation techniques. Unfortunately, there are significant differences between the economics of current wind farms and the economics of new wind farms entering a wind-dominated energy market. First, so far we have presumably been developing the best, most economic locations (i.e. those with high wind speeds at low altitudes near populations centers, but out of the direct view of influential people). Like prospecting for oil, we won't ever entirely exhaust wind power. Rather new wind farms will be relegated to increasingly substandard sites until they are no longer worth the effort. Second, wind is

---

[15]Here we are defining dense wind farms to have a power density of $1\,\text{W/m}^2$.

intermittent. It varies dramatically and unpredictably, but can be correlated over large areas. Like solar, this becomes a problem when it is responsible for a significant fraction of the power on an electricity grid. Fortunately, the wind tends to be stronger in the winter. This is because the equator does not experience seasons (so it stays roughly the same temperature), while the North and South Poles cool during winter. Hence, the equator-pole temperature difference is larger in winter (compared to summer), so the thermodynamic drive for the wind becomes more efficient. Importantly, this seasonal variation allows for synergy with solar, which is stronger in the summer. Because of this, we will see that the amount of wind energy (relative to human consumption) is a critical parameter in determining what energy options are sustainable. We will explore the consequences of intermittency further in Section 2.4.

2.3.6 *Biomass*

Seeds are super amazing devices. They contain a small amount of stored energy (just enough to get a plant started) and DNA instructions for self-assembly. They literally assemble themselves out of thin air. Even IKEA can't compete! First, a seed sends up a bright green shoot in search of sunlight. Presuming success, it can start a process you are likely familiar with: photosynthesis. Our newly born sapling takes sunlight, along with H_2O drawn up through its roots, and uses it to capture passing CO_2 molecules (i.e. carbon dioxide). The energy of the sunlight enables the plant to break apart CO_2 and H_2O to form O_2 and the carbohydrate CH_2O. The carbohydrate is added to make the plant imperceptibly larger and the O_2 goes off on its merry way (maybe to be breathed in by you).[16]

---

[16]The process of photosynthesis is considerably more complex than portrayed here. It includes a number of steps by which the carbon dioxide and water molecules are disassembled and the carbohydrate created. From these steps, we see that the oxygen in CH_2O comes from the CO_2, not the H_2O. Hence, most of the mass of a plant (i.e. the carbon and oxygen) come from the air, while only the hydrogen and a small amount of minerals are drawn from the ground. This

The critical feature of this process is that the plant needed solar energy to break apart the CO_2 molecule and assemble CH_2O. Hence, in theory we could get the CH_2O and O_2 back together and reverse the process to get the solar energy back. This is simple — it's called burning and it only requires a match to get things started. This is all there is to biomass. Use plants to capture and store solar energy, then burn them to run a steam cycle and generate electricity.[17] The upside is that we don't have to worry about intermittency because we can burn the plants on-demand. The difficulty arises in confronting the following question: Can biomass farms do any better than solar farms? Unfortunately, the answer currently appears to be "typically not."

Photosynthesis enables roughly 1% of incident solar energy to be stored as chemical energy in carbohydrates (though it is lower at high light intensity). To burn carbohydrates and produce electricity we must use a steam cycle, which reduces the efficiency by at least a factor of two. Thus, only 0.5% of the original solar energy will become electricity, which compares poorly against the 10% efficiencies typical of solar panels. From the average power density of sunlight ($200 \, \text{W/m}^2$), we see that powering our global energy needs with 0.5% efficient biomass process would require 20% of all land that isn't covered by ice. Moreover, this estimate is almost certainly too generous because it is optimistic about the efficiency of combustion and photosynthesis (e.g. the optimal biomass crops can't survive in many climates). Additionally, it doesn't account for the energy required to plant, farm, harvest, and transport crops, which, considering the low energy density of biomass, is substantial.

At a more practical sunlight-to-electricity efficiency of 0.1%, we would need to devote *all* of the Earth's ice-free land to biomass in order to satisfy our current energy needs. However, not all ice-free land can be cultivated. We use land for many other purposes

---

contradicts our conventional understanding of plants growing up from the ground. Instead, it is more accurate to picture them condensing out of air!

[17] Alternatively, the plants can be fermented to produce an oil (i.e. biofuel), which is then burnt.

(notably pastures for raising animals) and most environmentalists would agree that we've already taken more than our fair share. So planting biomass from sea to shining sea is neither practical nor desirable. A more achievable goal (but still ambitious) would be to double the current amount of cultivated land in order to devote 10% to biomass. This could be used to provide about 10% of our energy needs. Thus, unless we can genetically engineer significantly more efficient biomass crops or find an economic way to cultivate the open oceans, large-scale electricity generation from biomass doesn't look promising. While we should certainly take advantage of any biomass that comes as a by-product of other human activities, we should not expect it to provide a majority of our future energy needs.

2.3.7 *Fossil fuels*

The formation of fossil fuels is a long and convoluted process. It begins with plants converting sunlight into chemical biomass via photosynthesis (as discussed in the previous section).[18] The difference here is that, instead of immediately harvesting the plants to be burnt, the plants must be left to decompose in very specific *anaerobic* conditions. Anaerobic means "without oxygen," which is difficult to arrange considering it is a dominant constituent of the atmosphere. Normally, dead plants are broken down by hosts of microorganisms using chemical reactions that require oxygen as an input. Without oxygen, it becomes a much slower and less complete process because it must use an entirely different set of reactions.

Luckily for us, these anaerobic conditions do and did occur naturally, although only in specific ecosystems (such as bogs and the ocean floor). Still, even in ideal circumstances, the remains of anaerobic decomposition take ages to build up and then must experience intense pressures to convert them into energy-dense fossil fuels. For example, coal starts out as peat, which is the remains of terrestrial plants after anaerobic decomposition. Peat is formed at

---

[18] Petroleum and natural gas are generally formed from plankton in seas rather than terrestrial plants.

the Earth's surface and has a similar energy density to biomass, but it takes thousands of years to accumulate a layer just 1 m thick. To get coal, the peat must then be transported far underground to experience intense geological pressure. This pressure gradually increases the material's energy density as it transforms into lignite and then finally coal.

A very rough empirical estimate of the efficiency of this process can be done by comparing two timescales: humanity's fossil fuel consumption timescale and the Earth's fossil fuel formation timescale. In order to be sustainable, our consumption can be no quicker than formation. While there is considerable debate about exactly how long fossil fuel reserves can last, the general consensus is certainly no more than a few hundred years. Compare this against the hundred *million* years that it took our fossil fuel reserves to form. Hence, we see that we are a factor of a million from sustainability. Instead of providing nearly all our energy, they can only provide about a millionth of it. The minuscule efficiency of fossil fuels isn't particularly surprising though, when you consider the torturous and contingent nature of their formation.

This disparity in timescales is also at the heart of the climate change problem. The oceans currently contain 25 times more carbon than we could ever put into the air by burning fossil fuels. This makes sense because fossil fuels are so tricky to form that it would be surprising if they dominated the world's carbon budget. The problem isn't the quantity of carbon we're releasing, it's the speed. It takes the oceans thousands of years to mix water on the surface with water near the bottom. So, when we burn fossil fuels on the timescale of a century, the effects are unleashed on a thin layer of surface water, dramatically increasing its carbon content and saturating it's ability to remove carbon from the atmosphere. If we could just use fossil fuels a hundred times slower, we could "safely" burn all available deposits.[19]

However, there are methods to capture the carbon emitted by fossil fuels and prevent its emission into the atmosphere. This is

---

[19]Here, "safely" ignores all the mortality issues related to air pollution from burning fossil fuels.

known as carbon capture and storage (or CCS). By piping the captured CO_2 into oceans or underground geologic formations, the carbon can theoretically be kept out of the atmosphere. The primary question becomes whether CCS can be done cost-effectively and reliably. Currently, it is unclear whether underground storage, the most popular option, will hold the carbon for a long enough time. If the stored CO_2 were to leak out within a century, it could still have serious consequences on the climate. That being said, since coal generates roughly 40% of global electricity, retrofitting plants with CCS technology (which typically reduces their emissions by 85%) is an attractive short-term way to reduce global emissions. Nevertheless, since known coal reserves can only last for roughly 150 years (at the current level of consumption), we will need to develop alternative energy sources relatively quickly regardless of CCS.

2.3.8 Hydroelectric

Like billard balls, the individual water molecules in the oceans each have their own velocity. They are constantly colliding with one another, which causes them to change their speed and/or direction. It has been shown that, as long as collisions are frequent enough, the distribution of particle velocities will closely follow a bell curve, which is shown in Figure 2.3. The overall width of the bell curve is

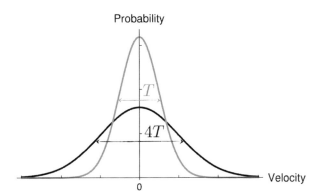

Figure 2.3: When collisions are frequent, the probability that a particle has a particular velocity follows a bell curve distribution with a width determined by the temperature. Here, we see the bell curve for two different fluids, the cold one with a temperature T that is four times smaller than the hot one.

what we call temperature, so increasing the temperature increases the number of particles with high speeds (though the overall shape of the curve stays the same).

When the Sun shines on the open oceans it increases the temperature of the water near the surface. This means that more of the water molecules (specifically the ones near the ends of the bell curve) have enough speed to break through the surface tension of the water and become gaseous water vapor. You probably already know that this process is called evaporation and it drives our planet's water cycle. Once an H_2O molecule gets into the air, it immediately rises like a helium balloon because it weighs less than the surrounding air molecules (which are N_2 and O_2). As the water molecule rises it encounters an increasingly cold upper atmosphere. Eventually, it is so cold that, if the molecule is absorbed by a droplet of liquid water, it no longer has enough speed to evaporate again. Hence, the water molecules can form clouds of liquid water droplets that are held aloft by slight updrafts. When the water droplets grow heavy enough (or the updrafts vanish), they fall to Earth as rain.

The average spot on Earth receives about a meter of rain each year. To determine how much energy this contains, we must calculate the gravitational energy released as all of this rainwater runs downhill from where it landed to the oceans. This can be estimated by multiplying the amount of rainwater by four quantities: the total land area of Earth, the density of water ($1,000$ kg/m^3), the strength of gravity (10 m/s^2), and the average land elevation (a bit less than a kilometer). The value this produces is roughly equal to the annual world energy consumption. However, like so many other renewable sources, we can't expect to harness it all. Over 50% of rainwater infiltrates into the ground or evaporates back into the atmosphere. Moreover, except in the case of waterfalls, the rainwater is continually losing energy due to friction with the land. Fortunately, the energy that remains is naturally concentrated in rivers, where hydroelectric power plants can efficiently extract it using water turbines. The most popular technique is to build a dam to create a large artificial lake upstream. This serves as both energy generation and energy storage, because the water can be released through the turbines on-demand. However, to minimize environmental impact, run-of-the-river hydroelectric power plants may be preferable. These plants extract energy from rivers as they flows by, without creating large upstream reservoirs. Though, because they have no storage, the energy provided by run-of-the-river plants is inherently intermittent, changing with the seasonal variation in river flow.

Optimistically, it looks like we may be able to install hydroelectric power plants on most of the world's rivers and generate roughly 10% of our energy needs.

2.3.9 Tidal

Tidal energy is very similar to hydroelectric energy. Both take advantage of a change in the elevation of water to generate electricity. For hydro, this elevation change is driven by solar heating and the water cycle, while tides are driven primarily by gravitational interactions with the Moon. Tidal forces arise because the Earth–Moon system hasn't quite reached its final equilibrium. Since gravitational attraction gets weaker with distance and the diameter of the Earth isn't tremendously smaller than the Earth–Moon distance, the Moon pulls on the near side of the Earth more than the far side. This means that, on the near side, the oceans get pulled slightly up off the Earth, while on the far side the Earth gets pulled slightly out from under the oceans. Hence, the oceans develop two slight accumulations of water, one at the location closest to the Moon and one at the location furthest from the Moon (see Figure 2.4 (right)). From this alone we wouldn't be able to extract any energy. The crucial fact is that these tidal bulges move relative to the continents because the Earth rotates so fast. In order to quantify the amount of tidal energy, we must first understand the motion of the Moon relative to the Earth's rotation.

In a few billion years, when gravity finally finishes its work, a day on Earth will be roughly 47 times longer than it is today and identical to the lunar month (i.e. the orbital period of the Moon). Until then, the mismatch between the rotation of the Earth and the orbit of the Moon means the Earth rotates much quicker than the two tidal bulges. This relative motion causes friction that slows down the Earth and speeds up the tidal bulges. Hence, the tidal

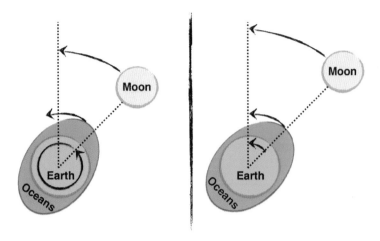

Figure 2.4: The present-day state (left) and the eventual equilibrium (right) of the tides. The Earth's rotation and the Moon's orbit are shown as viewed from the North Star. Nothing in this figure is to scale!

bulges move slightly ahead of the Moon and gravitationally tug it forwards. In response, the Moon speeds up and moves upwards to a larger orbit.[20] This process gradually equalizes the Earth day and lunar month, until the relative motion of the tides and the ocean floor is eliminated. In practice, the picture is more complex as the Sun also causes tides (though they are about half as strong) and the continents get in the way, messing up our pretty two-bulge cartoon.

While the Earth–Moon gravitational system contains a prodigious amount of energy (enough to supply our current needs for more than 100 million years), it does not appear easy to access. Though the energy reserve is enormous, we only have a small spigot to dispense it: gravitational interactions across 400,000 kilometers of empty space.

To estimate the amount of renewable tidal power that can be delivered by our gravitational spigot, it seems like we would have to measure the amount of friction between the oceans and land, a difficult task.[21] Instead, it is much simpler to measure changes in

---

[20] Note that this process increases both the kinetic and gravitational energy of the Moon. Hence, tidal power actually comes from slowing down the Earth's rotation.
[21] Tidal dissipation is mainly due to the sloshing of the oceans, rather than flexing of the solid body of the Earth. This dissipation primarily occurs at the floor of

the motion of the Earth and the Moon. During the Apollo missions, American astronauts positioned a number of mirrors on the surface of the Moon. Since we know the speed of light and have very accurate atomic clocks, we can bounce laser beams off these mirrors and precisely measure changes in the distance to the Moon. What we find is that the Moon is moving away from us at a rate of 4 centimeters per year. Additionally, our atomic clocks also tell us that the Earth's day is getting longer by 2 milliseconds per century. As explained in the "Estimating tidal power" Tech Box, these two measurements are enough to calculate that the total amount of tidal power is just a tenth of humanity's current consumption. A small spigot indeed!

TECH BOX: Estimating tidal power

To estimate the amount of tidal power, we must account for three stores of energy in the Earth–Moon gravitational system: the rotational kinetic energy of the Earth, the orbital kinetic energy of the Moon, and the gravitational potential energy of the Moon. The tides take rotational energy from the Earth and give it to the Moon's kinetic and gravitational energy. However, due to the sloshing of the ocean, some energy is lost in the process, which we can intercept and use to generate electricity. To calculate exactly how much, we must find the difference between the energy lost by the Earth and the energy gained by the Moon. Determining the energy lost by the Earth is easy — it can be directly calculated from the measured changed in the Earth's day (i.e. lengthens by 2 milliseconds per century). Similarly, the gravitational potential energy gained by the Moon can be found directly from the measured change

---

(Continued)

---

shallow seas like the North, Bering, and Yellow seas. Conceivably, through planet-scale engineering efforts (e.g. increasing the area of shallow seas, damming off oceans, etc.), we could significantly increase the total amount of dissipation and hence the available tidal power. However, it seems unlikely that this could be done economically.

(*Continued*)

> in the Earth–Moon distance (i.e. increases by 4 centimeters per year). Lastly, using conservation of angular momentum and both measurements, we can calculate the change in the Moon's orbital kinetic energy. With all our ducks in a row, we find that the amount of power that is dissipated in the Earth's oceans is approximately 10% of humanity's current total power consumption.

To harness the power of the tides, we typically use methods similar to hydropower. Like hydroelectric dams, we can build tidal dams, which capture high and low tides and use the elevation differences to run the water "downhill" through a turbine. This, of course, allows for energy storage, but the volumes of water enclosed by the dams must be enormous because the elevation change is usually only a few meters. Typically, damming off preexisting inlets (like the Bay of Fundy near Nova Scotia) is imagined. Alternatively, you can build water turbine farms in places where the tides create particularly fast currents. This works exactly like wind turbines and is almost identical to run-of-the-river hydroelectric plants.

Although there isn't a tremendous amount of tidal power available, we should certainly make use of it as it has a number of advantages. First, there are relatively few tidal power plants in the world, so there is room for development. Second, the technology is similar to hydroelectric and wind, so hopefully the economics are as well. Third, tidal power plants are unobtrusive as they are mostly (or entirely) underwater. Lastly, tidal power is either very predictable (water turbines) or it includes energy storage (tidal dams), so intermittency isn't as much of a problem.

2.3.10 *Wave*

The power carried by waves will be apparent to those reading this while standing chest-deep in the sea. The higher the waves are and the faster they rush to shore, the harder it will be for you to stay standing. Ocean waves are created by complex frictional interactions as the wind blows across the water's surface. Once waves are born, sustained winds can amplify and accelerate them until they reach land. Just like electromagnetic or seismic waves, these ceaseless oscillations can carry energy vast distances. And all the while they receive continued input from the wind.

If you look at a square of ocean, the wave energy it holds is determined by the height of the waves.[22] However, what we are most interested in is the total amount of power that arrives at a given section of the coast. This can be found by taking the energy per unit area and multiplying it by the wave velocity (i.e. how quickly our square moves to shore) and the length of the coastline.[23] Plugging in reasonable-sounding values (i.e. 1 m tall waves that arrive at a running speed of 10 m/s along the world's 10^5 kilometers of exposed shoreline), we find wave power to be 500 times less than wind and 10 times less than our current world power consumption. Admittedly, this is a very rough estimate. However, it is in keeping with more accurate and detailed analyses.

Why is wave power so much smaller than wind? The reason is buried in the details of the frictional coupling between wind and waves. Simply put, using one fluid to push another does not work well. When pushing one solid object with another, the bonds between all of the molecules force them to move in concert. Only a small number of movements are possible, so it is much easier to achieve the one you intend. In fluids, the freedom possessed by each individual molecule allows much more complex motion, namely turbulence. Turbulence takes nice orderly flows of air or water (or plasma!) and messes everything up. The flow distorts into swirls within swirls within swirls until the motion becomes heat. Essentially,

---

[22] More precisely, the energy per unit area is given by $\rho_w g h^2/8$, where $\rho_w \approx 1{,}000 \text{ kg/m}^3$ is the density of water, $g \approx 10 \text{ m/s}^2$ is the strength of gravity, and h is the wave height.

[23] Protip: To stand with ease amongst strong waves, try facing *along* the waves, instead of facing them head-on. Because people are wider than they are deep, this will reduce your own personal "coastline" and minimize the wave power your body receives. It's also sure to impress your friends at a beach party... especially if you explain it.

the vast majority of wind energy dissipates into heat at the water-air interface, rather than inducing ocean waves or currents.

The technology to harness wave power is relatively straightforward, but still in its infancy. Many designs exist, but none have been deployed at any significant scale. The simplest example is a single floating buoy, attached by cable to the sea floor. As the buoy rises and falls with waves, the motion is used to drive an electric generator. A more complex example, the Pelamis, is a long snake-like buoy that is designed to flex as waves pass by. This flexing drives the motion of hydraulic cylinders, which powers an electric generator.

Wave power has the potential to be useful in specific circumstances (most notably, remote islands). However, the amount of wave power suggests that it won't be particularly important in our global energy mix. Still, given that we currently don't utilize it at all, the only way is up.

2.4 Tackling Intermittency

In the previous sections, we have seen that several attractive energy sources are intrinsically intermittent. Most notably, solar and wind each contain enough power to be quite useful, but can vary significantly throughout the day, the week, and the seasons. Finding a cost-effective way to deal with the intrinsic variability of these sources turns out to be a challenge, but it is worth considerable attention.

The present-day electrical grid has very little capacity to store energy. Hence, at every moment, the electricity put on the grid by power plants must be removed by consumers with their air conditioners, toasters, and computers. At least for now, we don't have to worry about this, because grid operators are constantly adjusting the outputs of power plants to make sure everything exactly matches. This is accomplished with two types of electricity production. Baseload sources (like coal and nuclear fission) provide most of our electricity and maintain a constant, dependable electrical output. Dispatchable sources (typically natural gas plants and hydroelectric dams) generate less energy, but have an electrical output that can be quickly adjusted to match demand. This indicates that large-scale energy storage must be a difficult problem. Electric utilities choose to pay to build power plants that sit partially idle in order to avoid storing energy. In this section, we are looking for a way to replace both our baseload and dispatchable sources with sources that are intrinsically variable.

Figure 2.5 shows real data displaying the season-to-season and day-to-day variation in the output of Germany's solar and wind installations as well as the variation in power consumption. Germany is an ideal country to study because it has already invested heavily into renewable energy. In the first row of Figure 2.5 we see that electricity demand fluctuates relatively little throughout the day and the year (relative to wind and solar production). In fact, the most significant variation is the reduction in consumption on the weekends. This means that we should aim to combine wind and solar output to produce a fairly constant electricity output.

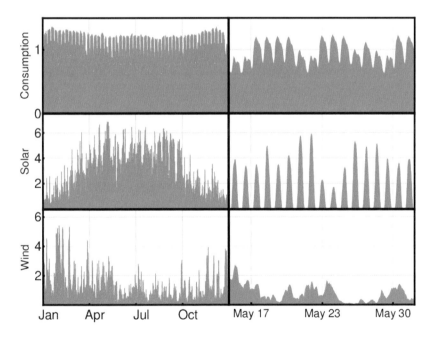

Figure 2.5: The power consumption, production of solar power, and production of wind power in Germany in 2016 (plotted on yearly and daily timescales) [10, 11]. All three quantities are scaled so that their yearly average is one.

The second row of Figure 2.5 shows Germany's solar generation, which displays several noteworthy features. First, because solar cells don't generate electricity at night and are usually somewhat misaligned with the Sun during the day, solar installations generate only a fraction of their maximum capacity. This is quantified by the capacity factor, which is the ratio of the energy generated compared to the maximum energy that could have been generated. Indeed, the capacity factor for solar in Figure 2.5 is about 15%. This means that, to produce a similar amount of electricity to a 1 GW nuclear fission power plant (which have typical capacity factors of 90%), you would need solar installations with a capacity of 6 GW. Second, we observe three dominant types of variation in the solar generation: daily, lulls, and seasonal.

The day-night cycle is obvious and predictable, but unavoidable. Solar installations don't generate any energy at night. This means that a hypothetical 100% solar energy supply must be able to store enough energy to make it from dusk till dawn, which constitutes about half of all electricity. We will see that storing half of the world's energy makes a 100% solar energy supply impractical.

Lulls, on the other hand, are less common and less predictable. They are usually caused by extended cloud cover, which quite literally comes out of the blue. In Figure 2.5, we see a typical lull, which starts on May 23 and lasts for three days. In these three days, the entire solar generation of Germany was reduced by half. Lulls are tricky because their appearance and severity are uncertain. While energy storage can help us outlast lulls, there are also other potential solutions. For example, since the power output isn't zero during lulls, we can build excess capacity. If Germany's solar energy generation was two times larger than what was actually needed, than lulls like the one between May 23–25 wouldn't be a problem. Similarly, linking geographically disparate regions using expansive electrical grids makes severe lulls less likely. After all, it can't be cloudy everywhere ... right? Lastly, since the likelihood of a lull decreases with its severity, a valid and economical option may be to concede at some point and reduce electricity *demand*. However, this could upset customers.

The last variation we observe in Figure 2.5 is the most concerning: the seasonal winter-summer variation. On average, the solar output in Germany is six times lower in December and January than it is in June and July. Though this seasonal variation is more predictable than lulls, we are still stuck with the same problem. Do we install six times the generating capacity we would need for summer? Do we build a truly prodigious energy storage system that can hold months worth of energy? What about building massive transmission lines between the northern and southern hemispheres? We will discuss these solutions in more detail shortly, but there may be a nice alternative: the wind.

The third row of Figure 2.5 shows the variation in Germany's wind power generation. Wind typically has a somewhat higher capacity factor than solar (here it is about 20%), but its output is much more chaotic. We see that, while wind power does not differ much between day and night, it still has lulls and seasonal variation. Crucially, wind power and solar power have opposite summer–winter variation. For instance, in Germany the wind is roughly three times as strong in the winter as in the summer. This raises the possibility of a very beneficial synergy with solar. By building the optimal combination of wind and solar, we can largely eliminate the energy that must be stored from summer to winter. This makes an insurmountable challenge start to look more feasible. However, since wind has less dramatic seasonal variation, we must have more of it to most completely cancel the larger variation in solar. For Germany, the optimal balance is estimated to be roughly 75% wind and 25% solar [12].

Through analyzing the variation in solar and wind, we have identified four strategies to address the problems posed by intermittent power sources:

- energy storage,
- demand management,

- expanding electrical grids, and/or
- excess generating capacity.

Here, we will briefly consider each in isolation, though the optimal solution almost certainly includes a complex combination of all four.

2.4.1 *Energy storage*

Even with the optimal balance of wind and solar, it is estimated that Germany would have to store 25–30% of all energy and have a storage capacity of 5–10% of their annual energy use [12]. Similar numbers have been estimated for the United States [13]. While this is less than it would have been for either wind or solar alone, it is still a tremendous challenge.

Currently, the cheapest energy storage method is pumped water storage, which uses electric pumps to move water up a hill when energy is abundant. Then, when energy is required, the water can be released back down through a hydroelectric plant to generate electricity. So we are converting electricity to the gravitational energy of water (which can be stored) and then back again. The overall efficiency of this process is around 75%, which is pretty good for energy storage. Unfortunately, it requires a lot of water and two big reservoirs, ideally at very different elevations. Even if we could use all of the liquid fresh water on Earth's surface, we would need to be able to lift it to the height of a ten story office building to store 5% of humanity's total annual energy needs.[24] Simply put, the amount of available water and the Earth's topology is not sufficient. This is not surprising because, in the hydroelectric section, we estimated that hydro could provide just 10% of humanity's annual energy needs. We would expect the storage capacity of such a system to be a small fraction of this. After all, much of this generation would be from

---

[24]The gravitational energy stored by water is proportional to both the elevation change and the mass of the water. Therefore, instead of lifting all the liquid fresh water on Earth's surface by ten stories, we could lift half of it by twenty stories or a third of it by thirty stories or...

run-of-the-river plants, which provide no storage. Furthermore, dams typically must release water many times throughout the year.

As with hydroelectric, we can also use geothermal to store energy. The basic idea is to heat water during times of energy excess and run it through underground pipes to transfer the heat to the ground. Later, when we need the energy back, we can run room-temperature water through the pipes to cool the rocks and bring the heat back to the surface. This is called underground thermal energy storage and it works because, as long as the temperature differences are small enough, the ground conducts heat very slowly (as we saw in the Geothermal section). We can put heat into the ground during the summer and almost all of it will still be there when winter rolls around. To estimate the land requirements for this, we will assume that we are heating up a 5 m thick layer of rock by 50°C. Using the heat capacity of rock, we find that every square meter can hold more than 10^8 J of energy.[25] This means that less than 0.1% of the Earth's land area is needed to store 5% of humanity's annual energy use! For this reason, underground thermal storage looks attractive, but only for certain applications. The temperature differences aren't large enough to efficiently generate electricity, but it could be used to heat buildings during the winter (which represents more than 10% of global energy use). This would effectively function as demand management, reducing the power that must be generated in the winter by wind. Since underground thermal storage has only been demonstrated at a small scale, there is considerable uncertainty about its eventual economics (especially in cities where the population density is high). Let's try it out!

Another storage solution, lithium-ion batteries, have made tremendous gains in recent years and now appear to be driving a revolution in electric cars. Batteries are attractive because you can typically get back over 80% of the electricity you originally put in. The Tesla Powerwall takes the lithium-ion battery technology

---

[25] The formula $\Delta E = C \Delta T$ gives the change in thermal energy (per unit volume) caused by heating a material. Here $C = 2 \times 10^6$ J/m^3/K is the specific heat capacity of granite and $\Delta T = 50°$C is the change in temperature.

developed for transportation and applies it to home energy storage. In 2017, the batteries in the Powerwall retailed for $5,500 and were guaranteed to store at least 37,800 kW-h of energy throughout their lifetime.[26] This corresponds to a cost of $0.15 to store a single kW-h of energy. Since it is estimated that 25% of energy will have to be stored, this would add just $0.04 to the average price for a kW-h of electricity. This incredibly low storage cost, enabled by mass production and economies of scale, is similar to the cost of generating electricity [14]. While this calculation is rough, it is consistent with more sophisticated analyses that found day-night battery storage would approximately double the cost of electricity of solar panels [6]. Hence, it looks like, with a bit more technical development, battery storage could achieve economic viability. Unfortunately, there is an important caveat: this is only true for *small-scale, day-night* energy storage. There are two significant roadblocks in scaling lithium-ion batteries up to *grid-scale, seasonal* energy storage.

The first is the sheer amount of raw material needed. Let's take lithium as an example.[27] In a typical lithium-ion battery, 1 gram of lithium is required to enable 4,000 Joules of energy to be stored. This means the capacity to store 5% of the world's annual energy needs would require a mind-boggling 6 billion tonnes of lithium. In the Fusion section, we estimated the world's potentially extractable terrestrial lithium to be just 0.03 billion tons, a factor of 200 less than we need. The world's biggest stash of lithium, the oceans, is about 30 times larger than we need, but this doesn't appear accessible at the necessary timescale or prices. The vast majority of lithium is deep in the ocean and will take 1,000 years to come to the surface. This wasn't a problem for fusion because the lithium was used gradually over millions of years. However, we can't wait this long to build

---

[26] The storage capacity of rechargable batteries gradually diminishes with time. The warranty on the Powerwall batteries states that their storage capacity will still be above 70% of its original value after an energy throughput of 37,800 kW-h or 10 years. After this, the batteries could still provide some useful storage, but about this Tesla makes no guarantees.

[27] Though cobalt resources are also particularly concerning.

the battery storage system needed for wind and solar to completely replace fossil fuels. Moreover, extracting lithium from seawater is still in the research and development phase and quite possibly will never attain the rock-bottom prices needed for economically competitive batteries.

The second roadblock is the timescale involved in seasonal energy storage. Lithium-ion batteries in electric cars or home energy storage systems are typically charged and discharged once a day or so. However, to store summertime solar energy for use in winter, we need to store energy for *months*, a very different challenge. This is because lithium ion batteries, even if they are just sitting on a shelf completely unused, only have a lifespan of around 10 years. This is problematic because, when we calculated the cost of energy storage above, we assumed that the batteries would be charged and discharged thousands of times. If we can only cycle a battery ten times during its life, then the cost of energy storage gets hundreds of times more expensive.

While pumped water storage, underground thermal energy storage, and chemical batteries are the most noteworthy, there are a host of other energy storage techniques. Unfortunately, each of them is currently lacking in some respect. Solar thermal energy, an alternative to photovoltaics, uses sunlight to directly heat a molten salt. The salt can then hold the energy for several days before being used to generate electricity. However, this requires a large centralized power station that currently has a cost of electricity several times greater than conventional methods [6]. Similarly, supercapacitors and superconducting magnetic energy systems, both of which directly store the electrons from electricity, also appear much too expensive to be practical anytime soon. Both hydrogen[28] and compressed air[29]

---

[28] Hydrogen storage uses energy to break H_2O into H_2 and O, which can be stored (though storing hydrogen turns out to be tricky too). You then can get the energy back when the H_2 and O are recombined.

[29] Compressed air storage works by squeezing pressurized air into a chamber (e.g. anything from small canisters to enormous underground caverns) and then extracting wind energy as the air is allowed to blow back out.

can be used to store energy, but they have low conversion efficiencies. This means that we would have to generate twice or three times as much energy as we would get back after storage.

2.4.2 Demand management

The second strategy is demand management, or adjusting the consumer side of the electrical grid to help make things balance. There are many activities that use energy, but are not particularly time sensitive. Refrigerators and freezers, for example, can usually be entirely turned off for more than an hour with no ill effect. Similarly, certain types of industrial manufacturing (e.g. water desalinization, cement mills, aluminum production) are also flexible. In many instances, demand management is functionally equivalent to energy storage. Both storing thermal energy underground and manipulating the fueling of batteries in electric cars can just as easily be thought of as demand management. In fact, if widely implemented, they would likely be the dominant source of flexible demand. In practice, demand management would likely be accomplished by utilities varying electricity prices according to instantaneous supply/demand. When the wind is still and clouds are out, prices would increase to discourage energy use. Smart equipment would be aware of the changing price and modify their energy use to save their owners money. However, unlike underground thermal storage, demand management does not appear to allow energy use to be shifted by months. So while it can play an important role in coping with daily variability and even lulls, its impact on seasonal variability is expected to be limited.[30]

2.4.3 Expanding electrical grids

Our third strategy, expanding and strengthening the electrical grid, is fairly intuitive. If we want to reduce the intermittency of wind and solar, we can use expansive, high-capacity electrical grids to enable

---

[30]To be fair, we don't have much experience with this. For example, how would energy demand restructure itself if electricity prices were substantially higher in the winter? It seems like it could lead to a lot of factories being shut down for months at a time.

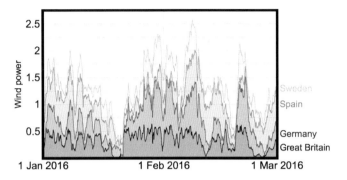

Figure 2.6: A stacked area plot showing the wind power generation in Great Britain, Germany, Spain, and Sweden [15]. The power from each country is scaled such that its yearly average is 0.25, then it is added on top of the data set below it.

large amounts of power to be transported between meteorologically diverse regions. Unfortunately, the weather tends to be somewhat correlated over vast geographic areas. Figure 2.6 shows the wind power generation from four countries that span most of Europe. We see that a continent-scale grid would help, but not as much as one might expect. The power still can drop below one-third of its yearly average value and this is during winter, when solar output is expected to be low.

Instead, what would really help with seasonal variation would be *intercontinental* electrical grids. After all, the equator gets the same amount of solar energy all year round and the Southern Hemisphere experiences summer when the North has winter. Thus, you could largely eliminate seasonal energy storage (though you would still need to address day-night variation and lulls). To this end, there have been several proposals to build solar power plants in North Africa and distribute the electricity throughout Europe using high-voltage, direct current transmission lines.[31] These power lines can carry vast amounts of electricity to long distances with minimal losses. Moreover, such transmission systems are already in use in China and Brazil (albeit at a smaller scale) and appear to have reasonable economics.

---

[31] For example, see the DESERTEC plan.

While intercontinental electrical grids appear to be one of the most technically feasible strategies to partially address the intermittency of renewables, they are one of the most politically challenging. Energy generation is one of the largest industries in all developed countries and this plan would require much of it to be outsourced to foreign lands. Imagine a country investing enormous sums of money to develop the industry and infrastructure of another country! Furthermore, it requires developed countries to fully abandon energy independence and rely on others for their energy supply. Solar-producing countries (and the countries crossed by transmission lines) would have the ability to disrupt the supply of electricity at a moments notice. Would the US really trust Columbia, Panama, Costa Rica, Nicaragua, Honduras, Guatemala, and Mexico in order to transport solar energy from South America?[32]

Regardless of whether we choose to *expand* electrical grids to support renewables, it is clear that we must *intensify* them. Current fossil fuel power plants are compact and can be located fairly close to where consumption occurs (e.g. cities). Sources of renewable energy are widely and unequally distributed according to the whims of Mother Nature. Hence, major sources of renewable energy generation are often in places that we never would have run transmission lines. Additionally, because of the increased uncertainty from renewable electricity generation, the electrical grid would have to be more flexible and have larger margins. This would force many existing lines to be upgraded to support more power.

2.4.4 Extra generating capacity

To illustrate the consequences of our final strategy, building extra generating capacity, let's take Figure 2.6 as an example. We see that, if we build exactly enough wind turbines to get by *on-average*, at certain times we will be short by a factor of three. Building three times what we need to get by *on-average* solves this. Unfortunately,

---

[32]Underwater high-voltage, direct current transmission lines are considerably more expensive.

the economics of building extra capacity in a capitalist system are of concern.[33] Building three wind turbines instead of one is not only three times as expensive, but also the outlook for the two extra turbines is more bleak and they can actually undercut the first. The two extra turbines will have to be built in less economically competitive locations and their output is largely redundant. Because of the correlated nature of the weather, the new turbines will tend to produce energy precisely when it is *least* valuable (i.e. at the same time as all the other turbines). So getting through the most severe lull requires us to build a wind turbine that is profitable at no other time of the year. Moreover, because it drives up the supply (at constant demand), it reduces electricity prices whenever wind is most productive. For this reason, without large-scale energy storage or demand management, the economics of an intermittent source get worse as it supplies an increasingly large fraction of the grid's energy.[34] Specifically, this problem becomes significant when the variability of a source causes it to regularly exceed the total demand across the entire electrical grid.

2.5 What is "Renewable"?

In light of our analyses of our energy sources and intermittency, it is useful to briefly consider the significance of renewable versus non-renewable energy sources. The term "renewable" is rarely given a precise definition, rather it is typically used to refer to all the sources we have discussed *except* for nuclear and fossil fuels. Strictly speaking, all that "renewable" implies about an energy source is that it is replenished. While this is certainly a good thing, it isn't sufficient to make an energy source attractive. What we really should be seeking are energy sources that are *sustainable* and *environmentally friendly*.

---

[33] Extra wind turbines would lower prices due to the economy of scale, but since enormous numbers are already being produced this effect looks to be quite small.
[34] If the electrical grid only has a handful of wind turbines, then electricity prices will be determined by the other energy sources. But if a large fraction of the electricity comes from wind, then the prices can vary substantially with the wind conditions.

Being renewable certainly helps with sustainability, but it doesn't guarantee that the renewable option is the better option. A renewable energy source doesn't necessarily contain more energy than a non-renewable one,[35] nor is it necessarily better for the environment. We saw that biomass can spur deforestation, hydroelectric dams create large artificial lakes, and that the manufacture of photovoltaic panels currently emits significant CO_2. The point is that we should take a comprehensive view in evaluating an energy solution and study the practical consequences of its implementation. All sources of energy have many pros and cons, which this chapter has explored. We should take this into consideration, instead of making judgments based solely on a single metric like renewability.

2.6 Outlook

In this chapter we have examined the dominant sources of energy available to us here on the surface of the Earth. Table 2.1 summarizes the main results. Judging from the energy hierarchy shown in Figure 2.1, we really don't see any surprises. As expected, fusion and solar dominate, though geothermal and tidal are maybe a bit smaller than we would have hoped. It turns out that, at the ripe old age of 4.5 billion years, the Earth's geothermal and tidal energy flows have slowed to a trickle. Ironically, the most remarkable feature of our planet's energy budget is probably that it contains so *little* fusion energy (compared to the Solar System as a whole).

In interpreting the content of this chapter, you should feel free to examine the assumptions, calculations, and logic and come to your own conclusions. We, in our interpretation, identify three broad methods for sustainable energy production:

- fusion,
- renewables, and/or
- nuclear fission.

---

[35]If all of the world's hydroelectric energy could be captured for the remaining 5 billion year lifespan of Earth, it would still be less than our current fusion energy reserves.

Table 2.1: Summary of the discussions in Section 2.3.

Source (fixed)	Quantity (at current use)	Intermittent?
D–T fusion	100,000,000 years	No
D–D fusion	10,000,000,000 years	No
Standard fission, land	100 years	No
Breeder fission, land	10,000 years	No
Standard fission, oceans	10,000 years	No
Breeder fission, oceans	1,000,000 years	No
Fossil fuels	100 years	No

Source (renewed)	Quantity (times current use)	Intermittent?
Geothermal	0.1	No
Solar	100	Usually
Wind	1	Yes
Biomass	0.1	No
Fossil fuels	0.000001	No
Hydroelectric	0.1	Usually not
Tidal	0.01	Maybe
Wave	0.01	Yes

Notes: The quantity of the fixed sources of energy is indicated by how long they could sustain our current world power consumption. The quantity of the renewable energy sources is indicated by the ratio of how much power we could capture to our current world power consumption.

These three plans can be termed "brains," "brawn," and "bravery," respectively.

The brains plan, fusion, requires the full intellectual and technological capacity of the human species to be marshaled against an unprecedented challenge. Success requires an enormous up-front investment in the form of research and development. Moreover, this must be done before the eventual economics of fusion energy are known. While the uncertainty is large, the consequences of success would be enormous. Fusion is by far the most attractive of our three energy plans, with the ability to operate quietly, reliably, and compactly with minimal environmental impact. Fusion power plants could replace existing fossil fuel plants with little change to our electrical infrastructure or our day-to-day lives.

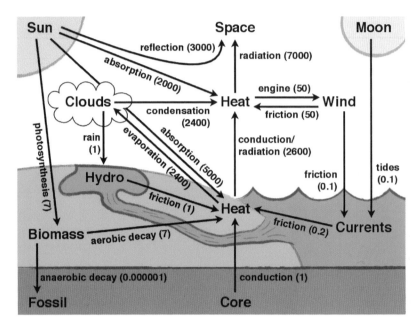

Figure 2.7: The magnitude of the flows of energy on our planet. All numbers are in units of our current global power consumption: 18 TW.

The brawn plan, renewables, requires using the full might of humanity's industrial capacity to forcibly divert substantial portions of the Earth's natural energy flows (as summarized in Figure 2.7). The diffuse nature of renewable energy necessitates the brute-force global deployment of millions of enormous man-made structures over much of the world's land area. Renewable energy's intermittency, combined with the difficulty of energy storage, motivates employing several distinct energy sources. While solar is abundant and has a high energy density (compared to other renewables), its seasonal variability currently prevents it from supplying our energy needs on its own. An elegant solution is to push wind power generation to its limits because it has the opposite seasonal variation. Moreover, hydroelectric and geothermal power can be maximized to provide energy storage, enabling an electricity supply that is more robust to the intrinsic lulls in solar and wind power. This strategy, along with an upgraded electrical grid, looks promising to provide roughly half of our energy. Pushing beyond this requires demand management, expanding electrical grids, and/or advances in energy storage.

Lastly, the bravery plan requires a global revolution in our relationship with nuclear fission power. Our collective mistrust of "nuclear" and "radiation" must be replaced by a public acceptance, not just to have it, but to have it in our own backyards. While global analyses of the history of fission power reveal it to be one of the safest sources of electricity,[36] this has not appeared to alleviate public fears. In the future, new types of breeder reactors must be developed, perfected, and deployed. When the inevitable mistakes occur, they must be met with a commitment to improve the industry, not abandon it. Finally, building fission power plants worldwide will increase the amount of nuclear material and the associated risk of a nuclear attack. For fission to thrive, society must be tolerant to this relentless low risk of catastrophic disaster.

At this point in time, we believe that these three plans are *not* options for us to choose from, but possibilities for us to aggressively pursue. All three include substantial uncertainty, not just about their economics, but their technical feasibility as well. Fortunately, the technologies that we must develop in pursuing them are broadly valuable, even if not implemented to generate electricity. Finally, it seems quite likely that the optimal solution will include a substantial role for more than one. The remainder of this book will exclusively explore the future of fusion energy, not because the others won't matter, but simply because it's what we can tell you about.

---

[36] A macabre tally of deaths per megawatt-hour reveals that wind and nuclear fission are historically the safest [16]. Hydroelectric is similar to nuclear fission in that the safety records of both are dominated by single accidents. Fission has Chernobyl, which directly killed roughly 50 workers and is expected to shorten the lives of approximately 4,000 people (due to an increased risk of cancer). The safety record of hydroelectric is marred by the Banqiao Dam collapse. The flood directly killed 20,000 people, while the resulting famine and epidemics killed over 100,000.

PART 2
THE BASICS

Chapter 3

Fundamentals of Fusion Energy

Imagine you're sitting at work one day and your boss hurriedly stops by your desk. He drops off some nuclei and asks you to go ahead and fuse them ... you know ... if it's not too much trouble. Presuming you knew little about fusion, what would you do?

Well, from things like Legos and welding, you know that to get two things to fuse you must bring them together and make them stick. So you take some nuclei in each hand and just sort of smush them together. This is, in fact, a particularly crude type of *cold* fusion[1] and it doesn't work. No matter how hard you try, at room temperature, the nuclei will never stick. They simply push off each other. Even if you get clever and push them together using the lattice structure of palladium metal, it still doesn't work. To understand why, we must learn about the fundamental properties of the nucleus.

3.1 The Nuclear Potential

The difficulty of nuclear fusion is that all nuclei have a positive charge (see Figure 3.1). They are exclusively composed of positively charged protons and neutral neutrons, which are collectively referred to as *nucleons*. The remaining constituents of the atom, the negatively charged electrons, are located far outside the nucleus, at distances

---

[1] Cold fusion refers to fusion at atmospheric temperatures, rather than stellar temperatures. It also goes by the more respectable-sounding name "Low Energy Nuclear Reactions."

Figure 3.1: A helium-4 nucleus, often referred to as an "alpha" particle for historical reasons. It is composed of two protons with positive electric charges and two neutrons with no charge. On this scale, the atom's two electrons would be far, far outside the picture (about the length of a football field away).

that are around 10,000 times greater than the radius of the nucleus. This means the presence of electrons has little effect on nuclear fusion. They are much too far away. Hence, pushing together two nuclei is like pushing together the north poles of two magnets. They repel each other and the repulsion gets stronger as they get closer. This is due to the force of electromagnetism. Fortunately, if the nuclei get close enough, a new attractive force called the *strong nuclear force* takes over and snaps them together.[2]

The details of this process are well summarized by what is called the *nuclear potential*,[3] which is shown in Figure 3.2. Each nucleus creates its own potential and experiences the potential of other nuclei. This is exactly analogous to the gravitational potential created by the Earth, which is shown in topographical maps. Just like it takes energy for you to go uphill in the Earth's gravitational potential, it takes energy for a nuclei to go "up" the potential of another nuclei. The only difference is that the nuclear potential is created by the combination of the strong nuclear and electromagnetic forces, instead of gravity.

---

[2] Strictly speaking, the force that binds nucleons together to form a nucleus is actually the *residual* strong force. The strong force is what holds the constituents of individual protons or neutrons together.
[3] Formally, the nuclear potential is the total amount of energy required to bring a nucleon in from very far away to a given location in space.

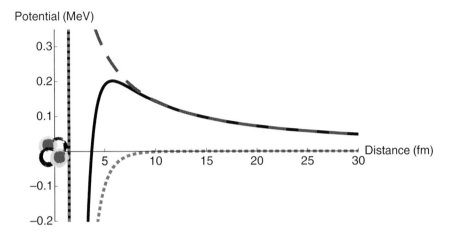

Figure 3.2: The nuclear potential of a helium-4 nucleus (solid), as seen by another charged particle. This is the sum of the potential from the strong nuclear (dotted) and electromagnetic (dashed) forces. Fusion occurs when a second particle approaches from large distances (i.e. from the right), climbs the potential barrier, and drops into the deep well.

Figure 3.2 shows that, when two nuclei are far apart, electromagnetism dominates and the particles repel each other. To see this, imagine rolling a ball in from the right-hand side of the figure. What would happen? It would slow down as it went up the "hill." Hence, repulsion! Of course, in reality there is no actual hill. You would just observe one nuclei slowing down as it approached the other. If the hill is tall enough, then the ball will come to a stop and then roll away, back down the hill. This is analogous to the two nuclei approaching each other, not having enough energy to overcome the electromagnetic repulsion, and flying back apart. This is called a *scattering collision*. However, if the ball has enough energy to get within a few femtometers,[4] we see that it can fall into the deep well. This well is created by the strong nuclear force, which dominates at short range and is generally attractive. When nucleons "fall" into this well, we call it *fusion*.

---

[4] One fm or *femtometer* is equal to 10^{-15} meters.

Figure 3.3: A mini-golf hole that is analogous to the nuclear potential (left) and what the hole would look like if it was to scale (right).

This means that fusion works a lot like a three-dimensional equivalent of the super-annoying mini-golf hole shown on the left of Figure 3.3. You have to get one nuclei really close to a second nuclei, so that it falls into the "hole" created by the second nuclei's strong nuclear force. Otherwise, it will simply roll back down the hill created by the electromagnetic force.

A particularly important feature for fusion is the height of the potential hill created by electromagnetism. This determines how fast the nuclei must be going to make it to the top and fall into the hole. Sadly, the hill is high, *very* high. The energy in the thermal motion of the molecules in air is 0.03 electron-volts (or eV), while the strongest chemical bond has an energy of roughly 11 eV (this new unit is explained in the "Units of nuclear energy" Basics Box).[5] These everyday energies are roughly a million times smaller than the peak shown in Figure 3.2. This is why cold fusion seems impossible.[6]

---

[5] The strongest chemical bond is a covalent bond between carbon and oxygen atoms.

[6] Fusion is so complex that we should never be too certain. We have just formulated what seems to be a robust argument against cold fusion, but an exception has already been discovered: *muon-catalyzed* fusion. A muon is an exotic particle that is like an electron, except roughly 200 times heavier. When an atom

The repulsion between the nuclei is a million times stronger than the tools we are using to push them together. It's like we're standing at the base of a mountain with a mini-golf putter, while the hole is at the summit (see the right-hand side of Figure 3.3). And we need a hole-in-one! Clearly we should employ more sophisticated tools to enable higher particle energies. In Section 3.3, we will do so, but first we must look at one last aspect of the nuclear potential: the depth of its well.

BASICS BOX: Units of nuclear energy

In the "Units of energy and power" Basics Box we saw that the usual unit of energy, the Joule, is too small to be convenient for the day-to-day energy use of humans. However, on the scale of individual atoms and nuclei, the Joule is much too large. Removing a nucleon from an atom only takes about 10^{-13} J. For this reason, scientists defined a new unit of energy: the electron-volt (abbreviated by eV). Formally, 1 eV is defined to be 1.602×10^{-19} J, i.e. the kinetic energy gained by an electron after being accelerated through a 1 Volt potential. However, the precise definition isn't too important. What is important is that the energy of chemical reactions is on the order of 1 eV. Nuclear reactions, on the other hand, are on the order of 1 million eV (abbreviated 1 megaelectron-volt or 1 MeV). The kiloelectron-volt (abbreviated keV), which is a 1,000 eV, lies in between.

Additionally, by using the Boltzmann constant of $k_B = 8.62 \times 10^{-8}$ keV per Kelvin, we can convert between the

(*Continued*)

---

is formed with a muon instead of an electron, the muon orbits are much closer to the nucleus, significantly lowering the peak of the nuclear potential. This enables fusion to occur at low temperatures. Unfortunately, muons require considerable energy to create and only live for a microsecond, so it looks impossible to use them to generate net fusion electricity. It's really strange stuff. You couldn't make it up.

(*Continued*)

average kinetic energy of particles in a fluid and an equivalent temperature. This is possible because the temperature signifies the width of the bell curve for the distribution of particle velocities (see Figure 2.3 on page 39 for a reminder). A fluid with very little average kinetic energy must be composed of particles that have a distribution of speeds with a very narrow width. As the average kinetic energy of the particles is increased, the bell curve widens, which means the temperature is getting larger. The Boltzmann constant is the appropriate conversion factor.

When we perform this conversion, we find incredibly large values! For example, 1 keV is roughly equal to 10,000,000 Kelvin, which is 10,000,000°C or 20,000,000°F. You might find this rule of thumb useful later because we will generally quote "temperatures" in keV, even though keV is really a unit of energy. Doing so is simpler than keeping track of all those zeros.

3.2 Binding Energy

In Figure 3.2, we saw that, at short distances, the nuclear potential drops precipitously. By zooming the vertical scale out, we get Figure 3.4. This shows that the nuclear potential well eventually bottoms out. Determining precisely how deep the well is or how far down a nucleon will settle is difficult. It brings in details of the strong nuclear force and quantum mechanics, which complicates our simple cartoon.[7] Fortunately, fully understanding these details is not necessary for our purposes. All we really need to know is that fusion

---

[7]In reality, the nuclear potential is more complex than described here. Particles don't have a single well-defined position or velocity, but instead a probability of having a range of different positions and velocities. Still, our cartoon is very useful because it is simpler to understand and accurately informs our intuition.

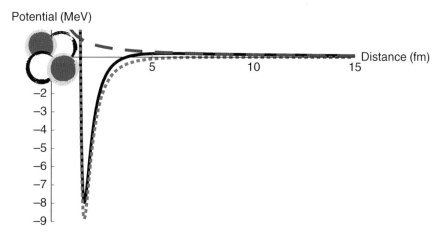

Figure 3.4: The depth of the nuclear potential well created by a helium-4 nucleus (solid), along with the contribution from the strong nuclear force (dotted) and the electromagnetic force (dashed).

is nucleons "falling" down into this potential well. The further they "fall," the more energy is released. To get a nucleon back out of the well, we must supply this same amount of energy. Importantly, as Figure 3.4 indicates, the well can be much deeper than the potential hill is high. Together, all of these complications determine what is called the *binding energy* of the nucleus and the energetics of fusion.

The binding energy is the amount of energy that is released in constructing the nucleus (i.e. dropping all nucleons one-by-one into the self-generated potential well). It is important because every nuclei has its own binding energy, determined by its unique internal structure. The most important quantity is the binding energy *per nucleon* because it indicates how tightly the average nucleon in the nucleus is held. Nuclei with a higher binding energy per nucleon are energetically favorable and tend to be more stable, since it takes more energy to pull them apart.

Figure 3.5 shows the binding energy per nucleon for a lot of common nuclei (the labeling of the nuclei is explained in the "Isotopic semantics" Basics Box). In order for a reaction to generate nuclear

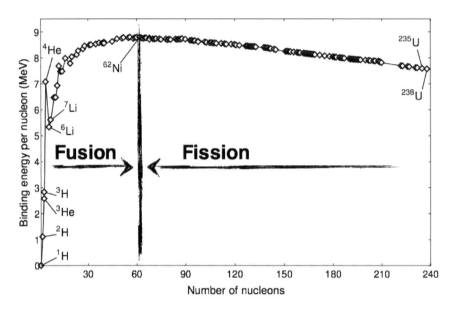

Figure 3.5: Binding energy per nucleon for the most abundant nuclei. Processes that move nuclei up in this plot will release energy.

energy, the inputs must have less binding energy than the outputs. In other words, we must convert nuclei into ones that are higher up in this diagram. This implies that, generally speaking, we want reactions that produce nuclei that are closer to having 60 nucleons. This clearly motivates both fusing nuclei with less than 30 nucleons each and fissioning nuclei with more than 120 nucleons. For example, Figure 3.5 tells us that fusing two hydrogen-2 nuclei to make one helium-4 nucleus will release 6 MeV per nucleon (i.e. 24 MeV in total).[8] Alternatively, we could split uranium-235 in half to generate about 1 MeV per nucleon (i.e. 235 MeV in total). But hold on, why are the most energetically favorable nuclei medium-sized? This is the

---

[8]This can be calculated by observing that the 4 nucleons in helium-4 each have roughly 7 MeV of binding energy, while the 4 nucleons in two hydrogen-2 nuclei each have 1 MeV. Hence, the difference between the initial and final states is (4 nucleons × 7 MeV/nucleon)−(4 nucleons × 1 MeV/nucleon) = 24 MeV or 6 MeV/nucleon.

final piece of the puzzle and can be understood from the nature of the electromagnetic and strong nuclear forces.

> **BASICS BOX: Isotopic semantics**
>
> In order to specify a nucleus, we must provide two numbers: the number of protons and the number of neutrons. The number of protons is indicated by the name and atomic number of the *element* (e.g. hydrogen has 1 proton, helium has 2, uranium has 92, etc.). This establishes nearly all of the *chemical* properties of the atom. This is because, since atoms are typically neutral, the number of protons alone determines the number of electrons (making the number of neutrons irrelevant). Hence, chemists can get by with the Periodic Table of the *Elements*.
>
> In contrast, neutrons are very important in nuclear physics as they affect the energetics of the nucleus. Hence, we must distinguish different *isotopes* of a given element. An isotope has the same number of protons, but a different number of neutrons. This is typically indicated by appending the total number of nucleons onto the end of the element name. So helium-3 has two protons and one neutron, while helium-4 has two protons and two neutrons. Alternatively, they can be abbreviated as ^{3}He and ^{4}He.
>
> Because the isotopes of hydrogen are so important, they also have their own special names. In addition to hydrogen-1, hydrogen-2, and hydrogen-3, we can call them protium (also known as normal hydrogen), deuterium, and tritium, respectively. Basically, it seems like the more important the nuclei is, the more names we have for it. The significance of isotopes makes the Periodic Table of the Elements insufficient for nuclear physics. Instead, all the isotopic information is "summarized" in an enormous diagram called the Table of the Nuclides (see Figure 3.6).

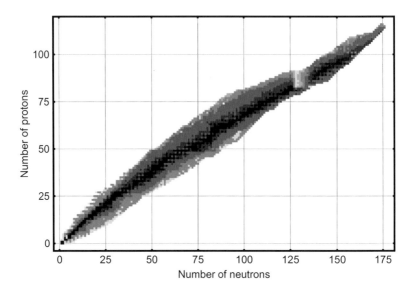

Figure 3.6: A very zoomed out view of the Table of the Nuclides. Each tiny pixel represents the stability of a different isotope (darker colors indicate more stability). Typically, each pixel in this table would also include detailed information about the method of decay and excited states.

Early in this chapter, we learned that the strong nuclear force is the only attractive force in the nucleus and it has a short range. In a small nucleus, adding a nucleon is strongly stabilizing because every nucleon now has another particle to attract it (see the top half of Figure 3.7). However, once you build a nucleus larger than about 3 femtometers, nucleons at one end of the nucleus are out of range of the strong nuclear force from the nucleons at the other end. This means that, in very large nuclei, adding a nucleon doesn't improve stability. In a very large nucleus, most of the nucleons are too far away to even notice the strong nuclear force from a new nucleon. Hence, the stabilizing effect of the strong nuclear force maxes out with increasing size.

Comparatively, the electromagnetic force has a long range. Even in large nuclei, every proton can still feel the electromagnetic repulsion of all the others. Since roughly half of nucleons are protons, this repulsion does not max out (see the bottom half of Figure 3.7). As nuclei are made bigger and bigger, the electromagnetic repulsion

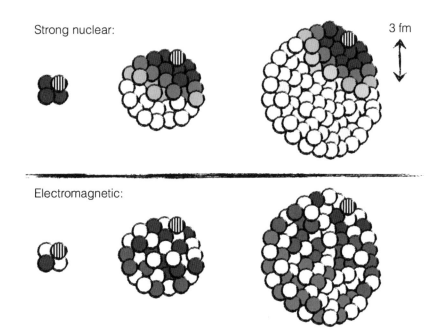

Figure 3.7: Feeling the forces: A new nucleon (marked by stripes) has just been added to each of three nuclei with increasing size. The darkness of the other nucleons indicates the degree to which it interacts with the new nucleon via the strong nuclear (top) or electromagnetic (bottom) force.

grows and grows until it overwhelms the maxed-out strong nuclear attraction, destabilizing the nucleus.[9]

---

[9] Why can't we just add a bunch of neutrons? This would increase the spacing between protons, which decreases the electromagnetic repulsion while keeping the strong nuclear attraction unchanged. To some degree, nuclei do this (note that the large, stable nuclei in Figure 3.6 have more neutrons than protons). However, the Pauli exclusion principle makes it energetically favorable for nuclei to have similar numbers of protons and neutrons. The Pauli exclusion principle states that two neutrons (or two protons or two electrons) cannot be in the same quantum state. This means that, as you add more of the same type of particle, they pile up. Hence, if we add a neutron to an already neutron-rich nucleus, it is forced to occupy a high energy quantum state, while a proton could occupy a much lower energy state.

This is why medium-sized nuclei are the most stable and energetically favorable. In small nuclei, adding a nucleon is stabilizing because of the attractive nature of the strong nuclear force. However, as more nucleons are added, this effect maxes out because of the nuclear force's short range. Since the electromagnetic repulsion has a long range, its repulsion does not max out with size, so it destabilizes large nuclei. This works a lot like going to the movies with a group of people. When the group is very small everyone can be friends with everyone else and you're still keen to meet new people. This is stabilizing because members of the group have a strong desire to overcome logistical hurdles to spend time with each other. However, when the group is sufficiently large, not everyone can know everyone else. So when some guy you've never met says he wants to watch *Gigli* instead of *Maid in Manhattan*, the group is likely to break apart![10]

Taking a step back, all this nuclear physics, and hence fusion, seems pretty strange. What we are saying is that, if we take two small nuclei and bring them progressively closer together, at some point they will suddenly gain a tremendous amount of kinetic energy. This energy comes from the potential energy of the nucleus, but why is it there to begin with? Why do the nuclear forces work in such a weird way? There really isn't a great answer. All we can say is that it's for the same reason that a ball rolling down hill gains kinetic energy: physics! Isn't it remarkable that the laws of physics allow us to extract so much energy from atoms in so many ways?

TECH BOX: $E = mc^2$

Albert Einstein's famous equation is strongly associated with nuclear power and, to some degree, this is appropriate. Two stationary hydrogen-2 nuclei weigh slightly more than

(Continued)

---

[10]Not that this has happened to me...

(*Continued*)

a single stationary helium-4 nucleus. When you take this mass difference m and multiply it by the speed of light squared c^2 (which is a big number), you find a tremendous amount of energy E. However, $E = mc^2$ is not a formula for *converting* mass into energy (or vice versa). Strictly speaking, it is a statement of the *equivalence* of mass and energy. In other words, if you weigh particles before fusion and then *immediately* after, they will still weigh the same. What!? Because the fusion products are moving very fast, they have a lot of kinetic energy, which corresponds to mass. It's only after the fusion products transfer away their kinetic energy and slow down that they weigh less than the input particles. In other words, what fusion really does is convert some of the rest mass from nuclear potential energy into the mass that corresponds to kinetic energy.

Moreover, the mass–energy equivalence is in no way unique to nuclear energy. If you weigh all the atoms in a piece of coal before and after burning (including the particles that get blown away as smoke, etc.) there will be a very small difference (assuming you have let the products slow down). Even riding in a hot air balloon will infinitesimally increase your mass because of the extra gravitational potential energy you possess at high altitude. In other words, just because nuclear energy involves larger amounts of mass and energy, doesn't mean that it has a special relationship with $E = mc^2$.

3.3 Fusion Cross-Section

Now that we understand why our attempts at cold fusion failed, it's time to get out the big guns. It seems simple enough. If the particles need more energy to climb the nuclear potential, let's give them more energy. So we build a high-tech particle accelerator and start firing nuclei at each other.[11] What we find is both frustrating and fascinating.

No matter what energy you set your particle accelerator to, the nuclei nearly always bounce off each other! We expected this for particles that were going too slow because they don't make it all the way up the potential hill. However, we observe that particles with too much energy don't fuse either. They overshoot and go flying past without falling into the potential well. It's like that annoying mini-golf hole in more ways than one. However, if we're persistent and use the right energy, every so often we're rewarded by a fused nuclei that comes flying out with tremendous energy. We've accomplished fusion!

Now comes the hard work. It isn't enough to accomplish fusion every once in a while. Running a particle accelerator consumes a lot of energy. If we want to generate net electricity (i.e. get more energy out than we put in), we must optimize our procedure to maximize fusion yield. So we sit down and systematically characterize the process of nuclear fusion. We take our accelerator and fire a whole bunch of different nuclei at one another. For each combination, we try a range of different energies. These measurements tell us a number of things. First, the energy with which the fusion products come flying back out tells us the energy each fusion reaction releases (we already saw this information in the binding energy curve shown in Figure 3.5). Second, they tell us the probability of each fusion reaction occurring. This probability is quantified by the fusion *cross-section*, which has units of area. Think of trying to get two tennis balls to hit in mid-air versus two basketballs. A reaction that has a larger cross-section is more likely to happen. However, while thinking of cross-sectional areas is intuitive, the reaction cross-sections are largely unrelated to the physical size of the nuclei involved. Rather, they have everything to do with the types of interactions between the particles and the stability properties of the products. Finally, our experiments allow us to measure how the fusion cross-section depends on the input energy. For electricity production, we are interested in reactions that

---

[11] Ernest Rutherford and his colleagues at the University of Cambridge did this in 1934.

release a lot of fusion energy and have a large fusion cross-section at low input energies.

From all of these measurements, a few patterns emerge. The first thing we notice is good news. Particles can still fuse, even when they don't have enough energy to fully climb the nuclear potential hill! This is because of a weird effect of quantum mechanics, which governs the behavior of these incredibly small objects. Because of the wave-like nature of matter and Heisenberg's Uncertainty Principle,[12] nuclei can actually "tunnel" through some portion of the nuclear potential. The effect of this is to lower the energies needed to achieve fusion by a factor of about twenty.

Second, we find that our simple picture of fusing two nuclei to form one is inaccurate. Fusion reactions are constrained to conserve both momentum and energy (factoring in the energy released by the reaction). In general, reactions producing a single particle have more trouble satisfying both these conditions, so they are very unlikely. Instead, we observe that two nuclei fuse and form two or more particles. Typically, fusion takes two similarly sized nuclei and produces one larger nuclei and a free nucleon (i.e. a proton or a neutron).

Our particle accelerator also reveals that nuclei with a lot of protons require much higher energies to fuse. This is not surprising because the positive charge of each additional proton increases the height of the electromagnetic potential hill. The high energies required for large nuclei make fusion prohibitively difficult. This is okay though, because we're not really keen to fuse these nuclei. The slope of the binding energy curve (shown in Figure 3.5) is steepest for super small nuclei (namely hydrogen and helium), so fusing them will increase their binding energy a lot and release the most energy.

Focusing our investigative efforts on only the smallest possible nuclei does *not* reveal broad, easily interpretable trends

---

[12]Without going into the details, the wave-like nature of matter and the Uncertainty Principle imply that the positions and velocities of subatomic particles have an inherent uncertainty to them. So a particle doesn't have a definite location, but rather it has probabilities of being in a range of locations. In effect, this means that a nucleus can teleport through the nuclear potential, even if it doesn't appear that it has enough energy to make it over the potential hill.

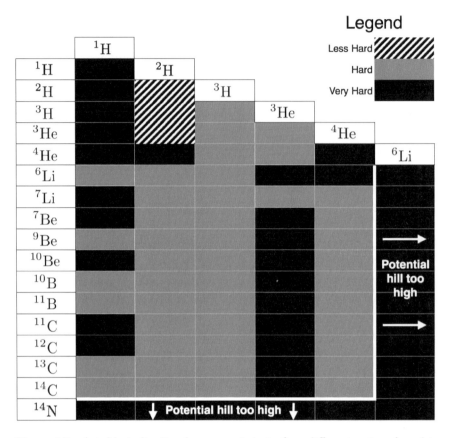

Figure 3.8: A table indicating how easy it is to fuse different pairs of nuclei. The easiest pairs have a large fusion cross-section that occurs at a relatively low energy.

(see Figure 3.8). Rather, everything depends on the precise details of which particles are combining and which particles they can produce. Though only one nucleon is different between hydrogen-3 and helium-3, these nuclei have significantly different behavior because there are only three nucleons in the nucleus. Nevertheless, through brute-force measurements, we discover one combination, deuterium–tritium (D–T), is substantially better than anything else. Apart from it, we find two other reactions that look reasonably promising: deuterium–deuterium (D–D) and deuterium–helium-3 (D–^{3}He). These are our fusion fuels.

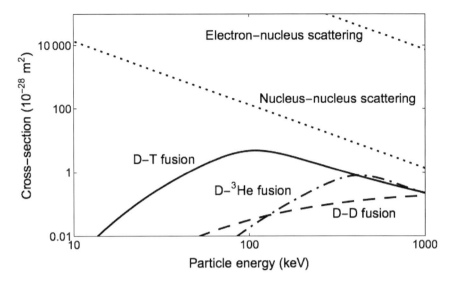

Figure 3.9: The largest fusion cross-sections (i.e. deuterium–tritium, deuterium–deuterium, and deuterium–helium-3) along with typical scattering cross-sections [17]. The precise identity of the nucleus (i.e. deuterium, tritium, or helium-3) doesn't change either of the scattering cross-sections much. Note that both axes are logarithmic.

Figure 3.9 shows the measured cross-sections for each of these fusion reactions, which indicates how likely it is for fusion to occur. The dominance of D–T fusion is immediately obvious. Its cross-section has a peak 5 times higher than the next best fusion reaction. Furthermore, the peak occurs at the remarkably modest energy of 100 keV (which is still 100 times greater than the average energy of particles in the Sun). Unfortunately, all the fusion cross-sections are dwarfed by the scattering cross-sections. Scattering (or colliding) means that the two particles approach each other and interact strongly, but ultimately bounce off without fusing. This alters the direction of both particles and allows for some amount of energy to be transferred, but nothing else. Figure 3.9 demonstrates that, even at incredibly high energies, nuclei are still much more likely to scatter than to fuse. This is an important fact to keep in mind when thinking about fusion: *nuclei are always much more likely to bounce off each other than to fuse.* It is also the fatal flaw of a fusion power plant based around a particle accelerator.

Imagine we maximize the fusion cross-section by giving every deuterium nuclei 100 keV of energy and firing it at a solid block of tritium. Because scattering is so much more likely, the deuterium will almost exclusively ricochet off of the particles in the material. Hence, instead of fusing, the deuterium will primarily transfer its energy to heat the solid material and/or scatter off out of the material. Either way, you will spend much more energy accelerating your particles than you will recover from the few that fuse.

While scattering collisions are inescapable, there is a fundamental strategy that can allow us to tolerate them: *confinement.* If the fuel is confined, then, by definition, it cannot scatter off out of the material. Moreover, it also solves the heating problem. As you accelerate deuterium nuclei into your confined region, they will initially heat the target material and cool down. But, after many collisions, eventually the system will reach something called thermal equilibrium (see the following Tech Box). At this point, the target material will be as hot as the deuterium and neither electron-nuclei nor nuclei-nuclei collisions will cause net energy transfer. Hence, scattering doesn't hurt us anymore. Let them scatter. We don't care because it no longer costs us energy.

TECH BOX: Thermal equilibrium

Thermal equilibrium is what results after many scattering collisions. If you inject a bunch of particles into a box and wait, they will rattle around exchanging energy for a bit until they reach thermal equilibrium. In this final state, individual particles still have different velocities, but the set of particles, taken together, will have a well-defined *distribution* of velocities. This distribution is called a bell curve and we saw it in Figure 2.4 on page 42. Once a set of particles has

(Continued)

(*Continued*)

> a bell curve distribution, collisions between them don't cause net energy transfer. The velocities of individual particles will continue to change, but the overall distribution of velocities will stay the same.
>
> Because collisions are so ubiquitous, most fluids follow a bell curve. This includes the water that you drink, the air that you breathe, and (for the most part) the deuterium that you ... would put in your terrestrial fusion power plant. In fact, strictly speaking, the concept of temperature is only defined when particles follow a bell curve distribution. The temperature is defined to be $T = (2/3)\, E/k_B$, where E is the average kinetic energy of the particles in the distribution and k_B is the Boltzmann constant. What this means is that a bell curve with a larger temperature will have more particles with higher energies.
>
> When several types of particles (e.g. deuterium nuclei, tritium nuclei, electrons) are in contact, they each still have their own distributions. However, for the entire system to be in thermal equilibrium, all the distributions must follow a bell curve and have the same temperature. Otherwise, the hotter types of particles will continue to transfer energy to the colder ones.

Confinement is a central issue of this book. How do you make fuel that is hotter than the Sun stay put? We will explore confinement (and the myriad of strategies to achieve it) at length in Chapter 4. But before we get ahead of ourselves, let's figure out exactly what it is that we are trying to confine.

3.4 Fusion Fuels

As Figure 3.9 shows, the D–T reaction has the highest cross-section at the lowest temperature, making it the ideal reaction for the first fusion power plants. The D–T reaction fuses deuterium and tritium nuclei, producing a helium-4 nucleus and a neutron. The reaction is

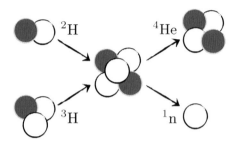

Figure 3.10: The D–T fusion reaction given by Equation (3.1).

summarized by

$$^2\text{H} + {}^3\text{H} \rightarrow {}^4\text{He} + {}^1\text{n} + 17.6 \text{ MeV} \qquad (3.1)$$

and is shown in Figure 3.10. Here, ^{1}n represents a free neutron and 17.6 MeV is the fusion energy released by the reaction. Enforcing conservation of energy and momentum requires that, because the neutron is so much lighter than the helium, it must carry most of the energy. Specifically, the neutron receives 14.1 MeV, while the helium-4 gets only 3.5 MeV.

However, at first glance, this reaction appears completely useless for a power plant. This is because tritium appears as an input and, if we zoom into our Table of the Nuclides, we find that its half-life is only 12 years. Since it decays away so quickly, Earth has no natural deposits of tritium. That isn't good. Normally, the absence of a fuel prohibits using that fuel as an energy source, but for tritium we make an exception. The D–T fusion reaction is so attractive that it makes sense to go to the trouble of creating the fuel using auxiliary nuclear reactions. It is a lot like IKEA furniture: it's clearly the best option, but there is "some assembly required." Though there are a few reactions that can *breed* tritium, the most promising uses lithium-6 and is given by[13]

$$^6\text{Li} + {}^1\text{n} \rightarrow {}^4\text{He} + {}^3\text{H} + 4.8 \text{ MeV}. \qquad (3.2)$$

---

[13]Note that breeding tritium from lithium is actually a fission reaction. Even though lithium-6 is on left side of the binding energy curve, this reaction still releases energy because helium-4 is incredibly stable.

The idea is to surround our D–T fusion reactions with lithium and use the free neutron from Equation (3.1) to make the tritium we need. Of course, we'll need some tritium at the beginning to start the first reactor up, but this can be produced in particle accelerators or nuclear fission reactors. Then, we can use the breeding reaction (along with techniques to multiply neutrons) to produce enough tritium to become self-sufficient. Hence, we see that the inputs to a D–T fusion power plant are deuterium and lithium, but we must learn to confine deuterium and tritium.

After D–T, the next hardest reaction is D–^{3}He and then D–D. The idea is to use D–T for first generation fusion power plants and later, after we become fusion experts, transition to a fuel that doesn't require tritium breeding. The D–^{3}He is given by

$$^2\text{H} + {}^3\text{He} \rightarrow {}^4\text{He} + {}^1\text{H} + 18.3 \text{ MeV}. \tag{3.3}$$

Unfortunately, helium-3 is rare here on Earth too and we see that the reaction doesn't produce a neutron, so we can't breed it. However, helium-3 is fairly abundant on the Moon, which might make it an ideal fuel for fusion-powered spacecraft!

Lastly, let's take a look at the D–D reaction. While the first two fusion reactions had only one possible outcome, this reaction actually has two branches. The two branches are equally likely and are given by

$$^2\text{H} + {}^2\text{H} \begin{matrix} \nearrow {}^3\text{H} + {}^1\text{H} + 4.0 \text{ MeV} \\ \searrow {}^3\text{He} + {}^1\text{n} + 3.3 \text{ MeV}. \end{matrix} \tag{3.4}$$

This reaction is attractive for terrestrial electricity production because, as we saw in Section 2.3.1, deuterium is extremely abundant here on Earth (no fuel breeding necessary). Additionally, the products of the D–D reaction are significantly different from the products of the other two reactions. Both the D–T and D–^{3}He reactions produce helium-4, which is an exceptionally tightly bound nuclei (see Figure 3.5). Because of its stability, helium-4 represents an end point of fusion. On the other hand, both branches of the D–D reaction produce nuclei that are inputs to the other fusion reactions. This means that, in the future, it is conceivable to operate what is called a *catalyzed* D–D fuel cycle, where we fuse the *products* of the

D–D reaction in subsequent reactions. Combining Equations (3.1), (3.3), and (3.4) has the net effect of

$$^2\text{H} + {}^2\text{H} + {}^2\text{H} \to {}^4\text{He} + {}^1\text{H} + {}^1\text{n} + 21.6\,\text{MeV}. \qquad (3.5)$$

Finally, if we were to surround our catalyzed D–D reactions with normal liquid water, we could breed a deuterium nucleus via the neutron capture reaction

$$^1\text{H} + {}^1\text{n} \to {}^2\text{H} + 2.2\,\text{MeV}. \qquad (3.6)$$

This enables what looks to be the ultimate nuclear fuel cycle:

$$^2\text{H} + {}^2\text{H} \to {}^4\text{He} + 23.8\,\text{MeV}. \qquad (3.7)$$

From the binding energy curve in Figure 3.5 it seems doubtful that we'll ever be able to do better than this. This, however, is a long way off. For the remainder of this book we will focus on the most near-term of goals: D–T fusion. We will see that it alone represents an incredible challenge.

3.5 Plasma

Due to the cross-sections shown in Figure 3.9, fusion requires incredible temperatures. Even for D–T, the fuel temperature must be at least 10 keV. But this is ten times hotter than the Sun! What does matter look like at these temperatures?

At these temperatures, all particles are in a gaseous state, flying around very quickly. So quickly in fact, that their kinetic energy is much greater than the binding energy holding electrons and nuclei together. Thus, when a particle collides with an atom, it has enough energy to knock the electron free, liberating it from its orbit. This process, knocking the electrons out of atoms, is called *ionization* and fusion-relevant temperatures ensure that all hydrogen and helium atoms are ionized. This leaves us with a brand-new state of matter: *plasma*. A "plasma" is a gas of free nuclei and free electrons, all flying around independently (see Figure 3.11). In a plasma, the free nuclei are called "ions" (hence the term ionization) and the free electrons are called ... well, they're still called electrons.

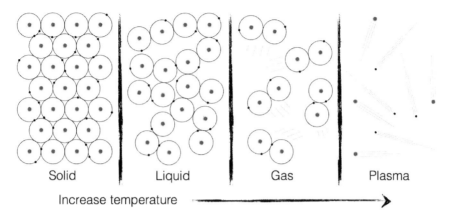

Figure 3.11: The four states of matter, where the small black dots are electrons, the larger gray dots are nuclei, and the black circles indicate the orbits of the electrons.

While fusion plasmas are definitely the rock stars of the plasma world,[14] plasmas appear all over the place. In the universe as a whole, plasma comprises over 99% of visible matter. Lightning, the solar wind, and fluorescent lights are all examples. Even fire can be considered to be a weakly ionized plasma (meaning that a substantial fraction of atoms have not been ionized). Initially, you might think that plasmas are simply very hot gases. However, their behavior is fundamentally different. It is far richer and more complex. As we will see, since the ions and electrons that comprise a plasma are charged, their motion produces long-ranged electric and magnetic forces. These forces affect the motion of nearby particles, which in turn alters the electric and magnetic forces that they create. From this, a whole range of exciting phenomena arise — phenomena that we must understand and contend with if we ever hope to confine a plasma.

---

[14] At least that's what fusion scientists think.

Chapter 4

Plasma Confinement

In the last chapter, we demonstrated the importance of confinement. Confinement allows us to tolerate the hundreds of scattering collisions that will happen (on average) before fusion occurs. This is where the real creativity of fusion enters. How do we design a system that keeps our fuel, a deuterium–tritium (D–T) plasma that is hotter than the Sun, in one place?

There have been many schemes proposed to confine a fusion plasma, but three in particular stand out. The first clearly works. It has a sustained history of success stretching back for billions of years and enables life as we know it. It is called gravitational confinement. If you have an astronomically large amount of fuel, it confines itself. Whenever a particle starts to wander off, its gravitational attraction to the rest of the fuel turns it around and brings it back. However, while this works in stars, we know that it is not practical for terrestrial fusion. Gravity is such a weak force that even the entire mass of the Earth would not provide sufficient attraction to confine a fusion plasma.

The second confinement scheme has also been confirmed to work. It was first demonstrated in 1952 and powers the most destructive weapon known to humans: the hydrogen bomb. This scheme is called inertial confinement. It works by imploding a sphere of fuel in on itself. As all the particles rush inwards, any that scatter and start to travel outwards are met by an onslaught of other particles. The idea is that, even if a particle tries to make a beeline out of the plasma, it will still fuse as long as there are enough other particles in its way. This uses the inertia of the fuel to provide *temporary* confinement. Of course, since there is no force continually acting on the sphere of fuel, it will quickly expand and the plasma will be lost. This can be tolerated if you can produce enough fusion energy quickly enough. Unfortunately, thus far, inertial confinement has only been achieved in explosions capable of destroying power plants, not powering them. Like gravitational confinement, scaling this technique down to something practical for energy generation proves difficult. Moreover, it has always required fission fuel in bombs, like uranium or plutonium, which endangers our global nuclear security (as we will learn in Chapter 10). A potentially attractive alternative for energy generation uses lasers, rather than fission, to implode the fusion fuel. This is an active area of research and will be discussed in Chapter 9.

In this chapter, we will explore a third confinement scheme, which is *nearing success*. It relies on a force stronger than gravity that can work in steady-state: electromagnetism. Before we start thinking about confinement schemes, we will identify a way to quantify confinement. This will help to give "nearing success" a precise meaning. Then, we will spend two brief sections reviewing electric and magnetic fields, which together comprise electromagnetism.[1] Experts should feel free to skip these two sections. Finally, we will be prepared to investigate a series of increasingly complex, but effective, electromagnetic confinement schemes.

4.1 Quantifying Confinement

In order to assess confinement schemes, it is important to have a metric for comparison. A useful parameter to this end is called the *energy confinement time*, which is represented by the symbol τ_E. This indicates how long, on average, it takes for energy to escape a

---

[1]Einstein's special theory of relativity reveals that electric and magnetic fields are actually two aspects of the same phenomenon, as indicated by the name "electromagnetism." By observing the motion of electric charges in a frame of reference moving at different speeds, you can smoothly convert between the two fields.

given confinement device. Strictly speaking, the confinement time τ_E is defined to be the total heat energy stored in the plasma divided by the amount of power continuously leaking out. So, if our confinement is really bad and energy leaks out really quickly, then the energy confinement time will be short and we will need a lot of heating power to keep the plasma hot. In other words, *better confinement equals a longer energy confinement time*. At the end of this chapter, we will see that, for D–T electromagnetic confinement schemes, an energy confinement time of $\tau_E = 1$ second is a pretty good goal to aim for.

We choose to use the *energy* confinement time, rather than the *particle* confinement time, because energy is the quantity that we ultimately care about. After all, we want to build a power plant that puts *energy* on the electrical grid. However, for the D–T fusion devices presented in this chapter, energy loss tends to be dominated by the kinetic energy carried by particles escaping the device. In other devices, electromagnetic radiation created by the particles can carry substantial energy. This can be important for devices that operate at higher temperatures (e.g. those using D–D or D–^{3}He fuel) and for devices with incredibly good particle confinement (e.g. the Sun).

4.2 Magnetic Fields

A magnetic field is a very curious thing. It can be thought of as a collection of magnetic field lines, which permeate all of space. Without mathematics, the magnetic field is easiest to understand by observing its effects on particles.[2] Particles without an electric charge, such as neutrons, don't interact at all with magnetic fields. Charged particles, on the other hand, have a fairly complicated response. The component of their motion *along* the magnetic field line is unaffected. So, a particle moving exactly parallel to a field

---

[2]The mathematical formula for the force on a particle due to a magnetic field $\boldsymbol{B}$ is $\boldsymbol{F} = q\boldsymbol{v} \times \boldsymbol{B}$, where q is the particle's electric charge and $\boldsymbol{v}$ is its velocity. Note that $\boldsymbol{F}$, $\boldsymbol{v}$, and $\boldsymbol{B}$ are vectors, so $\times$ is the cross product (not a normal multiplication).

line (in either direction) wouldn't notice any force. However, if a charged particle has a component of its motion *across* a magnetic field line, then it is deflected in the direction perpendicular to both the field line and the particle's velocity. This means that particles traveling exactly across magnetic field lines will have their trajectory bent into a circle. Note that the magnetic field never adds or takes away energy from a particle, it just changes its direction.

Most particles are moving partly along the field and partly across it, so their motion gets deflected into a helix (i.e. a combination of uninhibited motion along the field line and circular motion across it). The radius of this helix is known as the *gyroradius*, an abbreviation of "radius of gyration." Light particles that are moving slowly will have smaller gyroradii (i.e. tighter spirals). Moreover, increasing the strength of the magnetic field (indicated by B) will also tighten the spiral. The strength of a magnetic field is represented by the local spacing between the field lines that comprise the field. For example, the magnetic field on the right side of Figure 4.1 is twice as strong as the one on the left. This figure also shows example trajectories of ions and electrons in a uniform magnetic field. Note that the direction of the spiral is opposite for electrons and ions, but doesn't depend on which direction the particle is moving along the field.

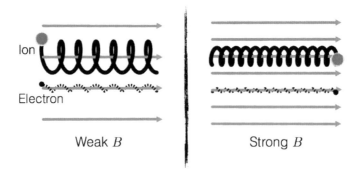

Figure 4.1: The helical motion of ions and electrons in a weak (left) and strong (right) uniform magnetic field B (denoted by the gray arrows). Particle motion *along* a magnetic field line (in either direction) is unaffected by the field. Motion *across* field lines is deflected into a circle, where the direction of gyration is determined by the particle's electric charge. A stronger magnetic field, a slower perpendicular speed, and a less massive particle (e.g. electrons) will all lead to a tighter spiral.

The nature of the magnetic force has profound implications for fusion. Figure 4.1 shows that charged particles tend to stick close to magnetic field lines, as long as the field is sufficiently strong. Since a plasma is just a collection of charged particles, this can give us considerable control over its behavior. In fact, the magnetic force directly provides confinement perpendicular to magnetic field lines, just not along them.

This looks promising, but we still need to know how to create magnetic fields. Once again, neutral particles don't play a role.[3] Rather, magnetic fields are only generated through the motion of charged particles (i.e. electric currents). Typically, we rely on the motion of electrons to create electric currents because they are much

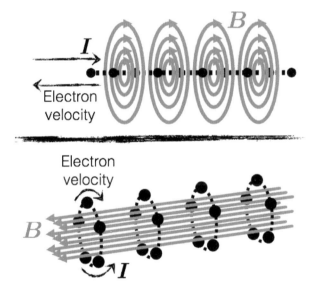

Figure 4.2: A line of electrons moving down a straight wire creates circular magnetic field lines (top), while electrons moving along circular wire loops create straight field lines (bottom). The magnetic field (gray arrows) is labeled by B and the electric current (black arrows) is labeled by I. Note that, because electrons have a negative charge, the electric current points in the direction exactly opposite to their motion.

---

[3]Sorry neutral particles...

lighter than ions, which makes them easier to move. The magnetic field created by a current is determined by the *right-hand rule*. If you point the thumb on your right hand in the direction of the current, then the resulting magnetic field will be a circle around the current in the direction that your fingers curl (see top half of Figure 4.2).

In day-to-day life, you are probably used to creating magnetic fields using fridge magnets, which are made from a material that has intrinsic electric currents. These are called *permanent magnets*, but their fields are not strong enough for our purposes. To generate very strong magnetic fields, we must organize many more electrons to all move long distances together. This is typically done using a metal wire, which can be bent into different shapes to create different magnetic field patterns. It is very common to bend the wire into a loop because, when you line many such loops up, it generates a straight magnetic field (see bottom half of Figure 4.2). These loops of wire are called *electromagnets* (also magnets or coils) and they can create exceptionally strong magnetic fields. Throughout the entirety of this book, when we refer to magnets we mean electromagnets (rather than permanent magnets).

4.3 Electric Fields

Fortunately for us, electric fields (represented by $\boldsymbol{E}$) are simpler to understand than magnetic fields. Electric fields push positively charged particles in the direction that the local electric field line points and negatively charged particles in the opposite direction (see Figure 4.3). Electric fields don't push neutral particles at all.[4]

You can think of electric fields as being generated in two different ways. The first way directly uses electric charges, which inherently create an electric field in proportion to their charge. A positive charge

---

[4]Sometimes it's nice just to be noticed...

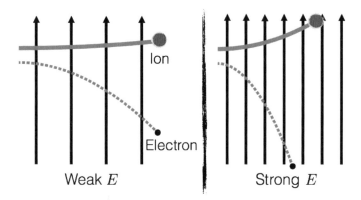

Figure 4.3: The motion of ions and electrons in a weak (left) and a strong (right) uniform electric field E (denoted by the black arrows). Particles enter from the left with purely horizontal motion and are accelerated either up or down by the electric field. Because electrons are much lighter than ions, they are easier for the electric field to push.

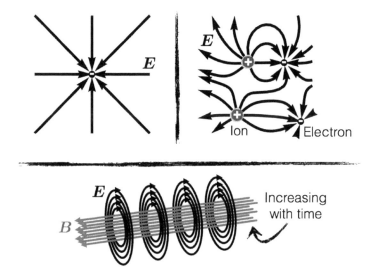

Figure 4.4: Electric fields (black arrows) generated by a single negative charge (top left), a set of positive and negative charges (top right), and a time-changing magnetic field (gray arrows) (bottom).

creates electric field lines that point directly outwards. Similarly, a negative charge creates field lines that point directly inwards (see the top left plot in Figure 4.4). More complex electric fields can be made by combining the effects of many charges (see the top right plot in Figure 4.4).

The second way to generate an electric field is through a time-changing magnetic field. As the magnetic field is changing, it creates an electric field that encircles itself (as shown in the bottom half of Figure 4.4). We have actually already come across this — electromagnetic induction! In Chapter 1, we learned that induction generators use a time-changing magnetic field to induce the motion of electrons and generate electricity. The force pushing the electrons along the wire is the electric field induced by the time-changing magnetic field.

4.4 Electrostatic Confinement

Now let's pick up where we left off at the end of Chapter 3. We've recognized the need to confine a hot fusion plasma, but how can we accomplish this using electric and magnetic fields? A simple strategy is shown by the electric field in the top left plot of Figure 4.4. If we can create a large accumulation of negative charge, then it will attract all the positively charged ions that we want to fuse. This would confine the ions exactly like gravity does, except better because the electric force is much stronger. Whenever an ion tries to leave, the electric field, which everywhere points inwards, would bring it back.

Unfortunately, creating an accumulation of negative charge is hard. We can't just stick a solid metal electrode into the center of the plasma and charge it with a bunch of electrons. After all, the entire point of the electrode is to attract the incredibly hot ions. Undoubtedly, many of the ions will collide with the electrode, simultaneously cooling the plasma and melting the electrode.[5]

---

[5]There have been attempts to make very minimalist electrodes in the form of spherical grids. This allows most ions to pass by without colliding. Unfortunately,

The confinement time of such a device would simply be the time it takes an ion to travel to the electrode. This is approximately $\tau_E = r/v$, where r is the radius of the plasma and v is the speed of a typical ion. A reasonable value for r would be a meter (i.e. the approximate size of a fission power plant reactor core), but v is a bit unclear. Earlier we learned that, in thermal equilibrium, the particles will have a distribution of velocities given by a bell curve. A natural choice is the most probable *speed* in the distribution v_{th}, which is called the thermal speed.[6] At a fusion-relevant temperature of 10 keV, this estimate gives a confinement time τ_E of about a microsecond, a million times less than our goal of 1 second. Also, after this you would have to rebuild your electrode. Not too good, huh?

A better solution would be to inject a bunch of extra electrons into the plasma. This would give the plasma a net negative electric charge, which would confine the ions. Unfortunately, plasmas do not tolerate net electric charge well. Because plasmas are composed of free charges moving at incredible speeds, they respond very quickly to cancel out any electric field. As soon as you start injecting electrons into the plasma, the inward electric field would force other electrons out of the plasma. This process ensures what is called *quasineutrality* (see the following Tech Box). Quasineutrality means that, at macroscopic distances, plasmas are electrically neutral (i.e. they have an equal amount of positive and negative charge). If a net electric charge is introduced into the plasma, the electrons and ions immediately respond to cancel out its effect as shown in Figure 4.5. Hence, using the plasma as the electrode won't improve the confinement time much, but at least it avoids destroying solid components.

---

the electrode needs to allow upwards of 99.99% of ions to pass, which doesn't look possible.

[6] Mathematically, the thermal speed is given by $v_{th} = \sqrt{2k_B T/m}$, where k_B is the Boltzmann constant, T is the temperature, and m is the mass of the particle.

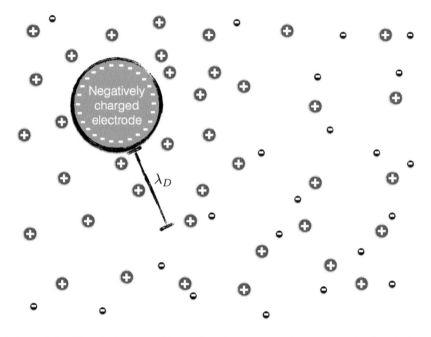

Figure 4.5: The electrons and ions of a plasma move in response to the electric field created by a solid electrode. Notice how the ions surround the electrode, while the electrons are pushed away. This behavior cancels the electric field at distances longer than the *Debye length* λ_D.

TECH BOX: Quasineutrality

While the plasma cancels out large-scale electric fields, the cancellation isn't exact. This is why we call it *quasi* neutrality, instead of neutrality.

First, while the electrons and ions respond fast, they don't respond instantly. The electrons, which are the lightest, set the fastest possible plasma response. Their response time is determined by the electron *plasma frequency*, represented by the symbol ω_{pe}. When presented with a new electric charge, electrons will oscillate around it at a frequency of $\omega_{pe} = \sqrt{ne^2/(\epsilon_0 m_e)}$.

(Continued)

(*Continued*)

Here, n is the plasma number density (i.e. number of electrons per unit volume), e is the magnitude of the electron's charge, ϵ_0 is the vacuum permittivity, and m_e is the mass of the electron. In a typical fusion plasma, this oscillation takes the electrons less than a nanosecond to complete.

Second, because the electrons and ions are moving so fast, they can't nestle right up against the charge and stay there. They invariably and continually overshoot a bit, meaning that very small-scale electric fields are possible in steady state. This is quantified by the *Debye length*, which is the distance over which the plasma cancels out electric fields. It is given by $\lambda_D = \sqrt{\epsilon_0 k_B T/(ne^2)}$, where k_B is the Boltzmann constant and T is the plasma temperature. In a typical fusion plasma, this distance is a fraction of a millimeter.

In general, plasmas are quick and effective at canceling out electric fields. However, because the cancellation isn't instantaneous or exact, the plasma can support the propagation of a wide variety of electromagnetic waves. For example, if you apply a time-changing electric field to the edge of the plasma, it will cause the motion of electrons, which can create an electric field further into the plasma, which causes the motion of other electrons, and so on. This perturbation (i.e. a wave) can propagate forwards until plasma conditions change sufficiently such that it is no longer supported. Later, we will see that these waves are very important for measuring properties of the plasma as well as heating it.

The fundamental problem with electrostatic confinement is that the material that creates the confinement necessarily attracts either the electrons or the ions. Moreover, using the plasma itself is impossible because the particles respond collectively to cancel electric fields and ensure quasineutrality. For these reasons, confinement with electric fields does not look promising.

4.5 Linear Magnetic Confinement

The bottom half of Figure 4.2 shows a strategy to overcome the limitations of electrostatic confinement. We see that a line of circular electromagnets creates a magnetic field that passes through each of the electromagnets' centers. This magnetic field can confine the plasma. More importantly, the objects that create the confinement (i.e. the electromagnets) are located *around* the confined region, not within it. Hence, we can separate the two areas and have solid material in one and a hot plasma in the other. Since particles are constrained to follow helices around the field lines, they are not attracted towards the solid magnets. They can happily gyrate away while sliding in one direction or the other along the field lines. They're stuck like beads on a string, away from any solid material.

But what about the ends? If our particles are stuck to the magnetic field lines like beads on a string, then what do we do about the ends of the string? Magnetic fields do not provide confinement in the direction parallel to the field lines, only the perpendicular directions. We could try ending them on a solid barrier, but it would melt and cool down the plasma just like the electrodes in electrostatic confinement. So what do we do?

Many people much smarter than myself have thought about this problem and they have come up with a number of clever solutions, one of which is the magnetic mirror (shown in Figure 4.6). The idea is to use what is called the *magnetic mirror force* to reflect particles as they approach the ends. A particle experiences the magnetic mirror force when it travels between two regions with different magnetic field strength. If the strength of the magnetic field is increasing, the speed of the particle along the field line gets converted into speed perpendicular to the field line. If the strength of the field is decreasing, the opposite conversion happens. The reason for this behavior is complicated (see the following Tech Box), but it is straightforward to see how this force could plug the ends of the device.

If we have regions of higher magnetic field near the ends, then particles will lose parallel motion as they approach them. If the field increases enough, the particles can lose all of their parallel motion and get reflected back towards the middle of the device. Unfortunately, the magnetic force acts on particles in proportion to their speed

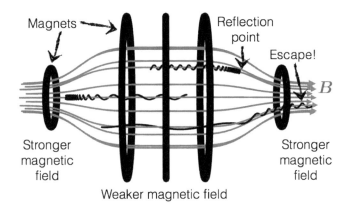

Figure 4.6: Three example trajectories (thin black lines) of ions in a basic magnetic mirror device. The top two particles, which are moving mostly perpendicular to the field line, get reflected by the regions of high magnetic field. Sadly, the particle at the bottom, which is moving mostly parallel, escapes the device.

perpendicular to the field line. So particles that are traveling mostly parallel to the field line won't experience a strong enough force to be reflected. This fact, that *particles with little perpendicular motion can escape magnetic mirrors*, appears to be a fatal flaw. Remember, the average particle will experience at least a hundred collisions before fusing and each collision changes the particle's parallel and perpendicular speeds. Hence, somewhere on the long road to fusion, there is a significant probability that a particle will scatter into a trajectory that escapes. Still this is an improvement over our electrostatic device. Instead of the confinement time being roughly the time it takes for a particle to cross the device, it is set by the time between particle collisions. For typical magnetic fusion parameters, this is on the order of a millisecond, a factor of 1000 improvement over electrostatic confinement[7]! Unfortunately, it is still much shorter than our goal of $\tau_E = 1$ second.

---

[7]A typical D–T magnetic fusion plasma has a temperature of $k_B T = 10$ keV and a density of 10^{20} particles per cubic meter. This information is enough to calculate the ion-ion collision time to be 15 milliseconds (i.e. a given ion will undergo a collision with any of the other ions in the plasma once every 15 milliseconds on average) and the electron–electron collision time to be 0.3 milliseconds. While we

---

TECH BOX: Magnetic mirror force

Imagine a particle that is spiraling around a field line, while also slowly moving along it into a region of higher magnetic field. Since the magnetic field strength is increasing, this implies that the neighboring field lines are getting closer together, gradually converging on the particle's field line. Since the particle has a non-zero radius of gyration it experiences these neighboring field lines, which aren't precisely parallel to each other (because they are converging). Since the magnetic force acts in the direction perpendicular to both the particle velocity and the *local* magnetic field, it will push the particle slightly backwards, while letting its radius of gyration expand a bit. As the particle continues into higher and higher magnetic fields, this small push continually converts more and more of the particle's parallel velocity into velocity perpendicular to its field line. If the magnetic field becomes sufficiently strong, it will entirely stop the particle's parallel motion. At this point, it will be making a circle instead of a helix. Still though, if this circular motion is occurring where the field lines are converging, then the particle will continue to be pushed. Hence, it will be accelerated back in the direction it came from as the entire process is reversed. This is the conceptual basis of the magnetic mirror and, if the change in magnetic field strength is large enough, it can reflect a large fraction of the particles in a plasma.

Research into magnetic mirrors started in the 1950s, along with fusion energy research as a whole. Scientists immediately recognized that magnetic mirrors have leaky ends. This prompted a series of experimental devices that explored ingenious, but ultimately unsuccessful solutions. Hence, they were stuck with a confinement

---

care most about keeping the ions hot, the electrons carry heat out of the plasma more quickly.

time similar to the time between collisions, which isn't long enough. To achieve fusion, we need to confine particles as they undergo hundreds of collisions. Though research continues to this day, magnetic mirrors fell out of favor in the mid-1980s. They were pushed aside by devices with better experimental performance ... by devices without ends to leak.

4.6 Combing a Hairy Ball

In studying linear magnetic confinement devices, we confronted the problem of end losses without much success. Simply put, confinement in two out of three directions is not enough. The natural next step is to seek magnetic configurations without ends. If the field lines wrap around on themselves and "bite their own tails," then there will be no ends from which to lose. However, as you will soon see, the set of possible magnetic field configurations is limited more by our creativity than by physics. In the three-dimensionality of our universe, magnetic field lines can bend and twist and weave and curl into all sorts of intricate patterns.[8] For this reason, a mathematical proof colloquially known as the *hairy ball theorem* is important. It tells us something that *isn't* possible — combing a hairy ball without creating a bald spot.

The hairy ball theorem implies that *it is impossible to wrap magnetic field lines around a sphere and completely cover it without at least one point having zero magnetic field.* The reason for the delightful name is the following analogy. Imagine a sphere that is covered in hair (i.e. a hairy ball). Now, try to comb the sphere such that the hair lies flat everywhere and covers every point. You can show that, for the hair to lie flat, there must be at least one bald spot (see Figure 4.7). For fusion, the hair is analogous to the magnetic field lines and the bald spot means that the magnetic field is zero. These zero points would cause problems for fusion, because no magnetic field means no confinement. Moreover, the hairy ball theorem doesn't just apply to spheres. It applies to any shape that is

---

[8]They can't diverge though.

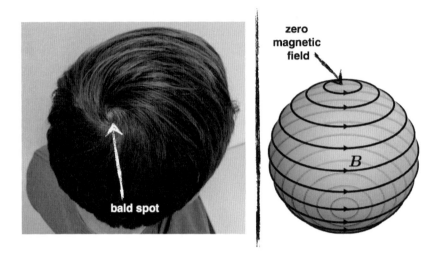

Figure 4.7: An attempt to actually comb a hairy Ball (left) and a sphere covered with magnetic field lines (right). In both cases, we see that zero points are unavoidable (at least that's what I tell myself).

topologically equivalent to a sphere. In English, this means any shape that you can create by deforming, squeezing, and/or stretching a spherical balloon.[9] These configurations are no good for fusion.

Instead, the hairy ball theorem motivates us to seek shapes that already have a hole in them, like a donut. Figure 4.8 demonstrates that a donut (and all topologically equivalent shapes) can be combed without issue. We see no zero points in the magnetic field, nor do we see magnetic field lines that would end on a solid surface. For these reasons, configurations with a hole possess the most promise for making fusion energy a practical reality. Collectively they are known as *toroidal* configurations because they are like a torus, the technical name for the shape of a donut.

4.7 Particle Drifts

Toroidal configurations have no ends or zero points, but that doesn't mean they can't have problems. If we take a look at the magnetic field

---

[9] No, you're not allowed to pop it!

Figure 4.8: A perfectly combed hairy donut! The magnetic field completely covers the surface of the donut without any zero points. We also see a trajectory (thin gray line) that we *hope* particles will follow — a helix around a field line that wraps around the device.

lines in Figure 4.8, nothing jumps out as being wrong. The particles should follow trajectories like the gray one, spiraling around and around the field line as they spiral around and around the device. It seems like they should be perfectly confined by their magnetic prison. If only things were so simple...

Now is the point where we must come clean with you. Previously, we *may* have skated over some details concerning the exact motion of particles in a magnetic field.[10] The reason for this was that, at the time, we had bigger issues to worry about. Who cares that particles don't exactly make a helix around one field line when they're about to slide off the end anyways? Well, now we're telling you. Particles aren't always perfectly stuck to field lines like beads on a string. They can stray sometimes. If the field lines are straight and uniform, then everything works nicely as advertised. However, as soon as you start bending or compressing parts of the magnetic field, the particles experience what are called *drifts*. Particle drifts, as the name suggests, cause particles to drift across magnetic field lines. There are

---

[10]Yes. We definitely did skate over details concerning the motion of particles in a magnetic field.

several different types of drifts, but here we only need to know three: the ∇B drift, the curvature drift, and the $E \times B$ drift.

The ∇B drift (pronounced "grad B") arises when the strength of the magnetic field has a gradient *across* the magnetic field line. In other words, a particle stuck to a given field line will experience the ∇B drift if the neighboring field lines have different magnetic field strengths. An example of this situation can be seen in Figure 4.9 (left). The physical origin of this drift isn't too hard to understand. As a particle gyrates around its field line, it passes through the neighboring field lines that have different magnetic field strengths. We know that a stronger magnetic field causes particles to have a tighter spiral (remember Figure 4.1). Hence, as the particle gyrates, it consistently curves more tightly on the side with the stronger field. This leads to the slow, but unavoidable cross-field drift shown in Figure 4.9 (left). Moreover, because electrons and ions spiral in opposite directions, they drift in opposite directions.

The curvature drift is caused by the curvature of a bent magnetic field line and is more complex. Any charged particle that is following a magnetic field line that is not a straight line will experience the curvature drift. Physically, it arises from inertia — the fact that particles travel in straight lines unless acted on by a force. Now, imagine a particle with no perpendicular motion that is exactly following a straight magnetic field. It will experience no force and continue in a straight line. However, if the field line starts to curve away from the particle's inertial trajectory, then the particle will gain a very small perpendicular velocity, which the magnetic field deflects. This deflection forces the particle to follow the curving field line, but also causes it to drift. The direction of this drift is perpendicular to both the magnetic field line and the direction the field line is curving. Figure 4.9 (right) shows an example of this. Because the magnetic field deflects ions and electrons in opposite directions, their curvature drifts will be in opposite directions.

The final drift that we need to know is the $E \times B$ drift (pronounced "E cross B"), which is particularly important. The $E \times B$ drift occurs when an electric field arises in the plasma and is, to some degree, "crossed" with the magnetic field. In other words, the $E \times B$ drift will occur as long as the electric and magnetic fields are not

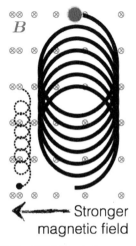
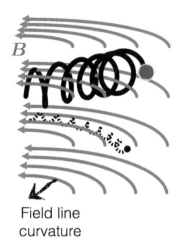

(Left) The ∇B drift, which occurs when a particle moves through a magnetic field with a strength that's not uniform in space. The drift direction is perpendicular to both the magnetic field (which here is pointing into the page) and the direction of its variation. Here the ions drift upwards and the electrons drift downwards.

(Right) The curvature drift, which occurs when a particle moves along a curved magnetic field line. The drift direction is perpendicular to both the magnetic field and the direction of its curvature. Here the ions drift upwards and the electrons drift downwards.

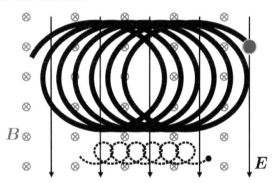

(Bottom) The $E \times B$ drift, which occurs when an electric field arises that is not parallel to the magnetic field. Electrons and ions drift in the *same* direction (i.e. here to the right), which is perpendicular to both the electric and magnetic fields.

Figure 4.9: Three particle drifts that are important in toroidal configurations.

exactly parallel to each other. The reason for the drift is similar to the ∇B drift. The electric field will accelerate a particle during one half of its gyration and decelerate it during the other. This will make the radius of gyration larger on one side than the other (because the particle is traveling faster on one of the sides). Importantly though, the electric field accelerates ions and electrons in opposite directions, which cancels out the fact that ions and electrons gyrate in opposite directions. This means that the $E \times B$ drift makes ions and electrons drift in the *same* direction.

Now that we've identified these three drifts,[11] let's understand how they apply to the simple torus shown in Figure 4.8. In short, they spell trouble. We can immediately see that all of the magnetic field lines are curved. From Figure 4.9 (right), we see that this will cause the ions to drift up and the electrons to drift down.

Additionally, the strength of the magnetic field also changes. This can be seen by imagining straight field lines that you then bend into a torus. The inside near the donut hole still has as many field lines as the outside, but it has less area. And remember, the closer the field lines are to each other, the stronger the field. Therefore, the magnetic field becomes stronger the closer you get to the hole of the donut. From Figure 4.9 (left), we see that the ∇B drift pushes ions up and electrons down. In other words, it reinforces the curvature drift.

Unfortunately, as soon as the ions start to drift up and the electrons start to drift down, you get a separation of charge. This, in turn, creates an electric field pointing downwards (i.e. pointing from the ions to the electrons), which causes an $E \times B$ drift. The $E \times B$ drift is outwards for both the ions and electrons, meaning the entire

---

[11] For the mathematically inclined, here are the expressions for each of the three drifts. The ∇B drift velocity is $\boldsymbol{v}_{\nabla B} = mv_\perp^2 \boldsymbol{B} \times \boldsymbol{\nabla} B / (2qB^3)$, the curvature drift velocity is $\boldsymbol{v}_\kappa = mv_\parallel^2 \boldsymbol{B} \times \boldsymbol{\kappa} / (qB^2)$, and the $E \times B$ drift velocity is $\boldsymbol{v}_E = \boldsymbol{E} \times \boldsymbol{B}/B^2$. Here, m is the particle's mass, $v_\parallel$ and $v_\perp$ are the particle's speed parallel and perpendicular to the magnetic field respectively, and $\boldsymbol{\kappa}$ is the field line curvature vector shown in Figure 4.9 (right). You can see explicitly that the ∇B and curvature drifts change sign with the charge q, while the $E \times B$ drift is unaffected.

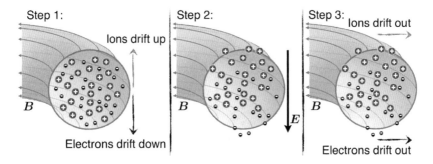

Figure 4.10: Recipe for failure. Step 1: The ∇B and curvature drifts cause a vertical separation of charge. Step 2: The separation of charge creates an electric field. Step 3: The electric field causes an outwards $E \times B$ drift of the whole plasma, ruining everything.

plasma moves until it hits something solid and loses confinement. These three innocuous particle drifts lead the plasma on a journey to its destruction (summarized in Figure 4.10).

While this entire process is complicated and takes quite a while to explain, Mother Nature executes it very quickly. Obviously, it depends on how nonuniform and curved the magnetic field is, but, for parameters typical of toroidal devices, it will demolish the plasma in a few microseconds! Interestingly, the ∇B and curvature drifts are fairly slow (less than a thousandth of the average particle speed), but, as soon as a small amount of charge separation occurs, the electric field rapidly moves the plasma. Back to the drawing board I suppose...

4.8 Toroidal Magnetic Confinement

The key to success for toroidal devices is to prevent charge separation. We can tolerate the ∇B and curvature drifts because they are fairly slow, but we can't allow them to create charge separation because this leads to an electric field and the fast $E \times B$ drift. Miraculously, preventing charge separation does turn out to be possible. It can be done by making the magnetic field lines wrap around the device in the short direction, in addition to wrapping around the device in the long direction (see the "Torus terminology" Basics Box and Figure 4.11).

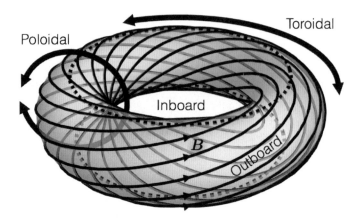

Figure 4.11: A torus with magnetic field lines that wrap around in both the toroidal and poloidal directions (a.k.a. the Thor and polar bear directions, respectively). The inboard side is the part that faces the donut hole, while the outboard side faces towards the outside. The boundaries between these two regions are denoted by the dotted lines on the top and bottom of the device.

BASICS BOX: Torus terminology

The technical name for the long direction around a torus is the *toroidal* direction, while the short direction is called the *poloidal* direction. Remembering these two terms will be important, because any discussion about a torus gets very confusing without them. To help, we've perfected the following mnemonic device: "*Thor* has *long* hair, while *polar* bears have *short* hair." See *Thor* the superhero has long glorious hair and his name sounds a bit like *tor*oidal, while *pol*ar bears have short hair and a name that starts like *pol*oidal.

Additionally, it is important to know the *inboard* side of the device from the *outboard* side (see Figure 4.11). The inboard side is the half of the torus that faces towards the donut hole, while the outboard side faces outwards, away from the rest of the torus. These two terms are easier to remember, so, if you want a mnemonic device, you're on your own. We spent all our creativity on the Thor one.

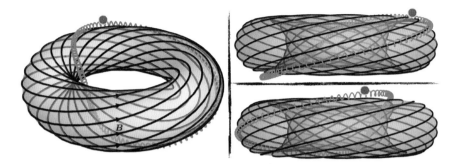

Figure 4.12: A bird's eye view (left) and two side views (right) of the same ion trajectory (gray line) in a magnetic field that wraps around in both the toroidal and poloidal directions. Importantly, after making one full poloidal circuit, the particle ends up where it began, even when the drifts are included.

To see why it is helpful to wrap the field lines around poloidally, we'll follow the path of a single ion (shown in Figure 4.12). Because of the ∇B and curvature drifts, this ion always drifts upwards in our torus. Hence, in the top half of the device it drifts out of the plasma, while in the bottom half it actually drifts further into the plasma. Now, it's important to remember that these drifts are much slower than the average velocity of the particles, typically by more than a factor of a thousand. This means that the ion is moving rapidly along the field line, around and around the device, while slowly moving upwards. So, if the field line wraps around in the short poloidal direction, then the particle will rapidly alternate between the top and the bottom of the device (as it slowly drifts upwards). Therefore, half of the time it is drifting *back into the device*, which cancels the time that it spends drifting out of the device. On average it doesn't drift at all! The same thing happens with electrons, except in the opposite directions. So, while all particles undergo small drifts back and forth across their field lines, there are no net drifts and no charge separation.[12]

---

[12] Strictly speaking, there is still a little charge separation, but the distance is smaller than the particle gyroradii.

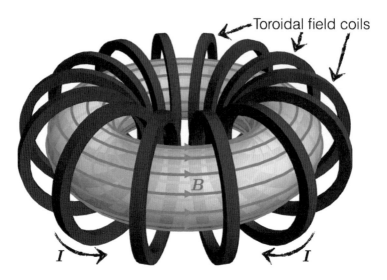

Figure 4.13: A device with fourteen magnet coils that create a purely toroidal magnetic field. The coils wrap around the poloidal direction, but are called *toroidal field coils* because of the field they create. The terminology can be a bit confusing because the toroidal field coils circulate current I poloidally around the plasma.

The question then becomes how to create such magnetic fields. Getting the field lines to wrap around in the toroidal direction is simple. We know that a line of circular electromagnets creates a straight magnetic field going through the center of each electromagnet. To create a toroidal field all we have to do is bend the line of electromagnets into a circle of electromagnets as shown in Figure 4.13. Easy stuff.

Creating the poloidal field is more of a challenge. The simplest thing to do would be to have a coil wrap around in the toroidal direction to create the poloidal field around it. Unfortunately, this coil would go precisely where the plasma is. Obviously, just like the electrodes in electrostatic confinement, we can't put a solid coil there without melting it. However, while plasmas can't tolerate a net electric *charge* (because they must be quasineutral), they can carry a net electric *current*. In fact, they can carry a current very well. Plasmas have an electrical conductivity that can exceed that

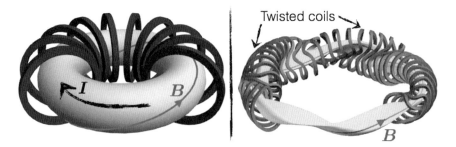

Figure 4.14: A single magnetic field line in a tokamak (left), which creates the poloidal magnetic field by running a toroidal current through the plasma, and a stellarator (right), which uses twisted coils that break the toroidal symmetry of the plasma. Some coils have been removed for viewing clarity.

of copper, the gold standard for electricity conduction.[13] This is the strategy pursued by a device called the *tokamak*, which is shown on the left side of Figure 4.14. Its curious name is a transliteration of the Russian name токамак, which itself is an acronym of тороидальная камера с магнитными катушками. This means "toroidal chamber with magnetic coils." Needless to say, tokamaks are a Russian invention. As we will learn, they rose to prominence in the 1960s due to the stunning experimental results of the T-3 device in Moscow. Tokamaks have flat toroidal field coils encircling the plasma (which generate the toroidal magnetic field) and a toroidal plasma current (which creates the much-needed poloidal magnetic field). Together the toroidal and poloidal fields combine to make helical field lines that wrap around the torus. Now, you may be wondering how the toroidal plasma current is created. This is a difficult technological challenge that we will explore next chapter. Nevertheless, the tokamak has achieved the best confinement of any terrestrial fusion device to date and will be the focus of much of this book.

A second strategy to create the poloidal field, devised by Princeton plasma physicist Lyman Spitzer in 1951, is pretty surprising. It can actually be entirely accomplished using external magnets! To be

---

[13]Plasmas also have a larger conductivity than gold, the gold standard for ... well the universe, I suppose.

honest, it's not at all obvious that this is even physically possible, but, indeed, it does work. These devices are called *stellarators*[14] and one is shown in Figure 4.14 (right). These fields are created by a bizarre set of twisted coils that directly force the magnetic field to wrap around poloidally. No plasma current is needed, but the mathematics of magnetic fields force us to break the symmetry of the plasma in the toroidal direction.[15]

Stellarators hold a lot of promise for fusion. In many ways, they are better suited for a power plant than tokamaks (primarily because they don't require a plasma current). We will explore the stellarator at length in Chapter 9, but here we will briefly describe their confinement properties relative to the tokamak. The main problem is that stellarators are very complicated.[16] Compare the two configurations in Figure 4.14. The tokamak plasma is smooth and svelte, while the stellarator plasma is twisted and peculiar. Fundamentally, the tokamak is a two-dimensional system (because it has symmetry in the toroidal direction), while the stellarator is three-dimensional. This has all sorts of profound consequences. Two of the most important ones relate to magnetic surfaces and fruit.[17]

4.9 Magnetic Surfaces

Imagine following a single magnetic field line around a toroidal device. In Figure 4.14, we see that, after one toroidal circuit of the device, the field line arrives back at its starting point, forming a closed loop. However, this is just one very particular example chosen

---

[14]Pretty cool name, right? The field of fusion is littered with fantastic names. Some of our personal favorites include the heliotron, the torsatron, the levitated dipole, the spheromak, the ELMO bumpy torus, and, of course, the Perhapsatron. Sadly, the Perhapsatron was not long for this world, so it was aptly named.

[15]Initially, all fusion research was classified as each country raced to "win." When everyone realized how hard fusion was and that winning wasn't going to come anytime soon, the research was declassified. For the most part, each country had been pursuing similar ideas, except for the stellarator. Lyman Spitzer was the only person in the world to come up with it. Looking at it, you can kinda see why...

[16]As fusion physicists, we don't say this lightly.

[17]This is not a typo.

to make a nice-looking picture. It need not be the case. If the poloidal magnetic field were weaker, the field line would angle differently and might require two or three or seven toroidal trips around the device before closing on itself. This balance between the magnetic field angling more toroidally or poloidally is quantified by the *safety factor* (represented by q). To put it precisely, the safety factor is *the number of times the field line goes around the donut toroidally for each time it goes around poloidally.* Hence, Figure 4.14 shows a $q = 1$ field line, while the purely toroidal field in Figure 4.13 would have $q = \infty$. In fact, as the second example in Figure 4.15 shows, q doesn't even have to be an integer.

In fact, because magnetic field lines are lines (i.e. they have no width), the vast majority of them never *exactly* close on themselves.[18]

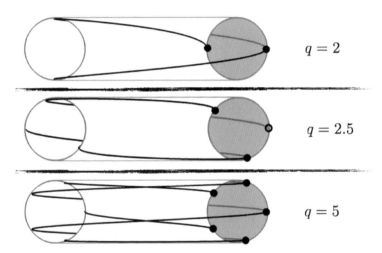

Figure 4.15: A front view of a single magnetic field line in three different tokamaks. Each field line has a different value of the safety factor, q. The top one wraps around the device twice toroidally each time it wraps around once poloidally. In the middle tokamak, the field line wraps around toroidally two and a half times during each poloidal circuit, while the bottom one wraps around five times toroidally.

---

[18] Formally, this is because there are many more irrational numbers than rational numbers.

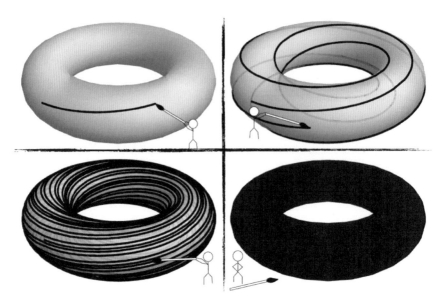

Figure 4.16: A person painting a magnetic field line will eventually color an entire surface, which is known as a magnetic surface.

They just wrap around the device infinitely in both directions (see Figure 4.16). We tend not to show these field lines in graphics because then we have to arbitrarily choose some point to stop drawing them. However, from a confinement standpoint, there is no problem with them, as long as the field lines stay on a single surface. The surfaces that are traced out by the field lines are called *magnetic surfaces* and they are very important for confining particles. Since particles are well-confined to field lines, if the field lines stay on a single surface, then so will the particles. Even if particles are free to explore the entire magnetic surface, it doesn't bring them any closer to the surface of the plasma. It's only by jumping to magnetic surfaces further out that particles come closer to escaping the device.

Tokamaks, because of their toroidal symmetry, are mathematically guaranteed to have nice, well-formed, nested magnetic surfaces[19] (like those shown in Figure 4.17 (top)). This is ideal to

---

[19]This can be seen by drawing one poloidal turn of a tokamak field line. It might come back to its starting point, but probably not. Regardless, because of toroidal

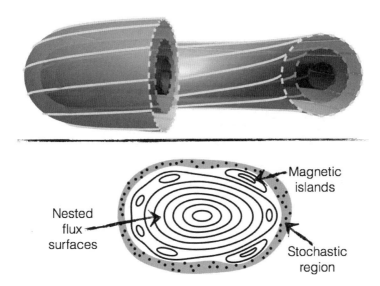

Figure 4.17: A cutaway view of three nested magnetic surfaces in a tokamak (top). Also shown is the poloidal cross-section of a poorly designed stellarator with stochastic regions and magnetic islands (bottom). Unlike the stellarator, each tokamak magnetic surface is guaranteed to be nicely nested within all the others (like those Russian nesting dolls).

maximize confinement. In contrast, there is no proof that magnetic fields in stellarators must form nested magnetic surfaces or even surfaces at all. In fact, they often don't. Instead, they get what are called *stochastic regions*, where a field line maps out a three-dimensional region instead of a two-dimensional surface. Additionally, *magnetic islands* can form, which are magnetic surfaces that aren't nested. Stochastic regions and magnetic islands are bad because they allow particles to get closer to freedom by traveling along the field line (which we know particles do very quickly and easily). Effectively, these two features act like a ladder in the game of Snakes and Ladders, giving the particles a short-cut in their escape route. This worsens the confinement, especially if the stochastic regions and/or

---

symmetry, we know that, if we move in the toroidal direction, all the field lines will look exactly the same. Hence, we can just take the field line that we have drawn and rotate it about the axis of toroidal symmetry and voilà! We have our magnetic surface!

magnetic islands span a significant fraction of the device. Hence, great care must be taken in stellarator design to ensure that the magnetic field forms tidy nested magnetic surfaces.

4.10 Bananas and Super-Bananas

In both tokamaks and stellarators, the slow cross-field drifts (i.e. the ∇B and curvature drifts) are neutralized by the fact that the magnetic field wraps around the device poloidally. This means that particles spend time in both the top and bottom halves of the device and drift inwards as often as they drift outwards. However, there is more to the story — not all particles travel the entire extent of the magnetic field lines.

When we studied the magnetic mirror, we learned of the existence of the magnetic mirror force, which reflects particles with large perpendicular velocity away from strong magnetic fields. Now combine this with what we learned when discussing the ∇B drift: the magnetic field is stronger on the inboard side of toroidal devices. Hence, we expect some fraction of particles to be reflected as they travel poloidally along a field line towards the higher field in the donut hole of the device. These particles are called *trapped particles* because they are trapped on the outboard side of the device. Additionally, their trajectories are called *banana orbits* for the reason shown in Figure 4.18 — they trace out a shape that looks like a banana.

Fortunately, in a tokamak this isn't too big of a deal. It's easy to see from the figure that, even though these trapped particles don't make it to the inboard side of the device, they still spend equal time in the top and bottom. Hence, their net drift is still zero and they stay close to their magnetic surface.[20]

In a stellarator, trapped particles pose a much more serious issue. Because the geometry is so much more complicated, the strength of the magnetic field goes up and down and up and down as particles travel along a field line. This means that, instead of particles only

---

[20] It can be shown mathematically that, if the magnetic configuration is toroidally symmetric, the trapped particles are guaranteed to have zero net outward drift.

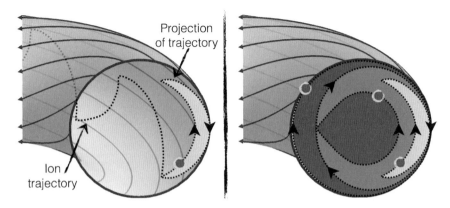

Figure 4.18: A banana orbit trajectory and its projection onto a poloidal plane (left) alongside three examples of particle orbits projected onto the poloidal plane (right). From innermost to outermost they are a typical banana, the so-called WFB (world's fattest banana),[21] and a particle that isn't trapped. Note that these trajectories are averaged over the particle gyration and the width of the bananas are exaggerated somewhat for viewing clarity (existing experiments typically have a radius larger than 20 banana widths or so).

getting trapped on the outboard side of the device, they can get trapped in all sorts of places. And many of these places are localized to the top or bottom of the device. Therefore, their cross-field drifts don't cancel, so they can escape the entire device. These unconfined trapped particle trajectories are called *super-banana orbits* because of how important they are.

This would seem to doom the stellarator, just like it doomed the magnetic mirror. Instead of leaking particles out of the ends, stellarators leak particles wherever super-bananas occur. Fortunately, there are many different magnetic field shapes for stellarators and some have fewer super-bananas than others. Just look at Figure 4.14 (right) and think about all of the other ways that you could bend those coils. In fact, the original stellarators were shaped like a Figure 8 (see Figure 4.19). As mentioned earlier, the possibilities are more limited by our creativity than anything else. Initially, this was

---

[21]This isn't a joke ... at least not anymore. "World's fattest banana" is actually a term that appears in academic journal articles. For example, see Ref. [18].

Figure 4.19: Professor Lyman Spitzer standing next to the first stellarator, which had a figure 8 shape.

actually a big problem. The sheer complexity of these configurations stymied researchers, setting back the development of stellarators by decades. Only in the last 20 years have researchers been able to attack the problem systematically. Now, powerful supercomputers are able to quickly compute millions of particle trajectories in order to evaluate and optimize the performance of a proposed stellarator. This process has identified magnetic configurations that are almost entirely super-banana-free, have nested magnetic surfaces, and can be created by coils that are reasonable to construct. Therefore, researchers are hopeful that these new "optimized" stellarators will have much improved confinement.

4.11 MHD Stability

This is pretty remarkable! After a long journey past many proposed confinement devices, we've finally identified one that perfectly confines individual particles: the tokamak.[22] If you dump an ion or electron into a tokamak, it will spiral around the magnetic field lines (and drift a bit in and out), but always stay close to its magnetic surface. Theoretically, individual particles have an infinite confinement time. In practice though, we must be careful — interactions between particles can make things less rosy. Because the electrons and ions in the plasma create their own long-range electric and magnetic fields, they can conspire together to break free from their magnetic cage. When there is just one particle in our tokamak, this isn't a problem, but, to maximize fusion power, we want to confine as much fuel as possible. Unfortunately, if we try to confine too much, the sheer force of the plasma will completely overwhelm the external fields we're using to restrain it.

In order to investigate this balance, early fusion researchers devised a simple model of the plasma known as *magnetohydrodynamics* or MHD for short. Instead of treating the plasma as a set of individual particles with their own positions and velocities, MHD makes several drastic approximations in order to treat the plasma as an electrically charged fluid.[23] In other words, it basically treats the plasma like water, except with an electric charge such that it generates and responds to electromagnetic fields. While, this treatment is usually *not* accurate, it is appealing because it is governed by equations that can be solved without needing powerful computers. Surprisingly though, MHD turned out to work well for certain applications — most notably predicting conditions that will lead to large-scale bulk plasma instability.

For example, MHD calculations told early fusion researchers that tokamaks with a safety factor q that is *less* than about 2 would be unstable. The fundamental cause of this instability is the plasma

---

[22] We've also identified a confinement strategy that *almost* perfectly confines single particles: stellarators.

[23] Formally, MHD assumes that collisions between particles happen very often, so that the distribution of particle velocities always exactly follows a bell curve. Moreover, it assumes the gyroradius is infinitely small, which means it doesn't include the physics of particle gyration nor banana orbits.

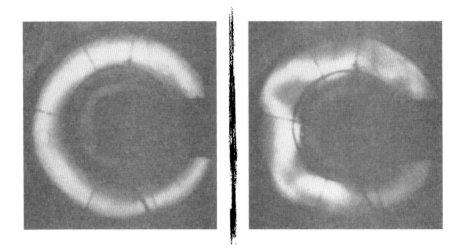

Figure 4.20: A top view of an early toroidal confinement device before the plasma starts to kink (left) and after (right). Unlike tokamaks, this device had no toroidal field coils, so there was no toroidal magnetic field to hold the plasma current in place.

current, which has a tendency to kink as shown in Figure 4.20.[24] However, if the toroidal magnetic field is strong enough, then it can resist the kink and keep the plasma current flowing normally. A low value of q corresponds to magnetic field lines that are angled very poloidally, which means the plasma current is strong compared to the external toroidal field coils. What MHD tells us is that, if we try to increase the plasma current so much that the safety factor drops below 2, the current will overwhelm the toroidal field and kink until the plasma slams into the nearest solid component and destroys itself. In fact, this instability is the reason the safety factor is called the safety factor.[25] Similarly, MHD gives us limits on other quantities

---

[24]Any electric current, straight or curved, has this same tendency to kink. In a normal copper wire, this instability is stabilized by the stiffness of the solid material (i.e. it takes energy to bend a wire). Plasmas, on the other hand, bend freely.

[25]In the next chapter, we will learn of the ZETA device, a proto-tokamak that was plagued by plasma instability because it had too low a safety factor.

like the product of plasma density and temperature (i.e. the plasma pressure).

Because of all the simplifications of MHD, it is unable to accurately predict a lot of plasma behavior. Yet, it is often useful for indicating conditions that make the plasma wildly unstable. While the precise details of these instabilities vary significantly, they typically lead to the destruction of the plasma within a millisecond. Hence, MHD stability is a necessary condition for good confinement that we must keep in mind. In Chapter 8, when we design a tokamak power plant, we will find that some of these MHD limits have important consequences for the economics of fusion.

4.12 Classical and Neoclassical Transport

In the very earliest days of fusion research, scientists hoped that, once single particle confinement and MHD stability was achieved, particle and energy transport would be *classical*. This refers to transport caused by scattering collisions as particles gyrate around magnetic field lines. We know that most of the time, when particles approach each other they bounce off, changing their velocity at random. Because gyration allows particles to stray from their field line by a gyroradius, a collision can cause a particle to jump to a new magnetic surface that is a gyroradius away (see Figure 4.21 (left)). However, since the direction of the jump is random, sometimes this would move particles further into the device. Still, this process does cause the transport of particles (and their energy) through what is called *random walk diffusion* (see the following Tech Box).

TECH BOX: Random walk diffusion

To understand random walk diffusion, imagine that you're at a swimming pool, standing in the middle of a diving board. Now flip a coin. If it's heads take a step forward and if it's tails take a step backwards. Then flip the coin again and take

(Continued)

(*Continued*)

another step. Repeat again and again. With each step there is a 50% chance that you'll move towards your starting point (partially canceling your earlier steps), but there is also a 50% chance you'll get further away. What is certain is that, if you take enough steps, you'll eventually fall off one of the ends (hopefully it's the one over the water!). Moreover, you'll fall off sooner (on average) if you take bigger steps or if you take less time flipping the coin between steps. When the mathematics of this situation are analyzed, we find that, after a time t, there is a 62% chance that you'll be more than a distance

$$x = \sqrt{\frac{(\Delta x)^2}{4\Delta t}t} \tag{4.1}$$

away from your starting location. Here x is the total net distance traveled, t is the total elapsed time, Δx is the distance traveled per step, and Δt is the time between steps. We see that, as the time t increases, a particle becomes more likely to be increasingly far away. Hence, a random walk process will transport particles out of the device, but it does so much slower than the usual directed motion.

We can use Equation (4.1) to estimate the confinement time in a tokamak. If we set x to be the distance from the center to the edge of the plasma, then t will be a typical time it takes for a particle to escape. As an illustration, we'll assume a plasma with a radius of 1 meter and take values common in existing magnetic fusion devices. For classical diffusion, we know that Δx is the ion gyroradius (roughly 1 centimeter) and Δt is the ion-ion collisional time (roughly 10 milliseconds).[26] Solving for t, we find the confinement time to be a few minutes.

---

[26]In classical transport, ion-ion collisions are the dominant effect, even though electron-electron collisions happen more often. This is because the ion gyroradius

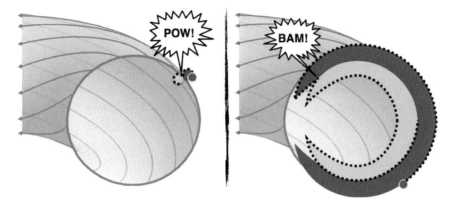

Figure 4.21: The trajectory of an ion (dotted line) as it undergoes a single collision. In classical transport (left), particle gyration enables this collision to move the ion to a new magnetic surface that is a gyroradius away. In neoclassical transport (right), banana orbits enable this collision to move the ion to a new magnetic surface that is the width of a banana orbit away. Note that both trajectories are projected onto the poloidal plane and the neoclassical one is averaged over the particle gyration.

If we start with a dense clump of particles in the center of our tokamak, collisions and random walk diffusion will cause them to travel outwards. Eventually, after many collisions, particles will start to reach the outermost magnetic surface and escape the device. Using the mathematics of random walks, we can estimate how long this will take in order to determine the confinement time (as is done in the above Tech Box). What we find is a confinement time of a few minutes, much longer than our goal of 1 second! Unfortunately, as you can probably guess, this estimate assuming classical transport is considerably optimistic.

Researchers quickly realized the importance of trapped particles and the banana orbits they make. In classical transport, we noted that gyration causes particles to stray by a gyroradius, so collisions

---

is much larger than the electron gyroradius and the confinement time depends on $(\Delta x)^2$.

can enable a particle to make a gyroradius-sized jump closer to freedom. Now, we remember that drifts create banana orbits, which typically carry particles more than a gyroradius away from their magnetic surface. This leads to so-called *neoclassical transport*. It's the same as classical transport, except, instead of using the gyroradius as the distance the particles jump, we must use the width of a banana orbit (see Figure 4.21 (right)). This hastens the random walk diffusion. After accounting for a host of complications,[27] we find the confinement time from neoclassical transport to be about a second. While this is substantially worse than the classical confinement time, it's still long enough for a viable power plant.

4.13 Turbulent Transport

What really blindsided fusion researchers was plasma turbulence. In the earliest days of fusion, MHD instabilities ended the plasma discharge before turbulence had time to appear. However, as the community became better at making fusion plasmas that lasted longer, turbulence started to rear its ugly head. Nowadays, turbulence is known to be the determining factor for the energy confinement time of tokamaks. It, more than anything else, is the reason that we don't have fusion power plants already. We've said that confinement is a central issue for fusion and now we'll say that turbulence is the central issue for confinement.

Plasma turbulence is like atmospheric turbulence (which you experience in an airplane) or the turbulence that occurs when you stir cold milk into hot coffee. It is characterized by swirling, chaotic fluctuations in the properties of the fluid (e.g. density, flow velocity, temperature, etc.) and appears when the driving forces behind the turbulence become stronger than viscous forces.[28] In other words, viscous fluids like syrup are very hard to make turbulent,

---

[27]The complications include the fact that trapped particles only represent a fraction of all particles and they have a shorter effective collision time.

[28]You may have heard of the Reynolds number before, which quantifies the balance between the kinetic forces driving turbulence and the viscous forces opposing it.

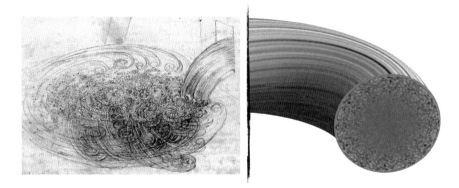

Figure 4.22: Leonardo da Vinci's 16th century illustration of water turbulence (left) and the turbulent density fluctuations in a tokamak plasma (right). Black indicates an accumulation of ions and white indicates an accumulation of electrons, so electric fields point from black to white regions. Notice that the turbulence is very elongated along the magnetic field lines and narrow across them.

rather, they tend to remain smooth and steady. Plasma, however, is the opposite of syrup. Collisions between particles in a plasma are much less common than in syrup (or water), so it has little viscosity and is prone to turbulence at the slightest provocation. Common drives for turbulence are spatial variations in the density and temperature of the fluid. Hence, fusion plasmas have the mother of all turbulent drives. After all, we want the center to be hotter than the Sun and the edge to be cool enough that it won't melt solid material. Moreover, we want this to occur in the space of a couple of meters.

Turbulence isn't a new problem (see Leonardo da Vinci's study in Figure 4.22). Understanding its behavior is important for designing cars, predicting weather, building bridges, and much more. The famous Nobel prize-winning physicist Richard Feynman described it as the most important unsolved problem of classical physics. Another great physicist, Horace Lamb, is reported to have said, "I am an old man now, and when I die and go to Heaven there are two matters on which I hope for enlightenment. One is quantum electrodynamics and the other is the turbulent motion of fluids. And about the former

I am really rather optimistic." In other words, turbulence is both incredibly important and incredibly challenging.[29]

However, plasma turbulence in tokamaks isn't your normal every-day turbulence. This is turbulence of electrically charged matter in the presence of a strong magnetic field. This means that, because turbulence causes different fluctuations in the densities of ions and electrons, it creates fluctuating electric fields. These small-scale fluctuations are called *turbulent eddies* and they are the dominant mechanism enabling particles to escape confinement (see Figure 4.23). The electric field of an eddy causes an $E \times B$ drift that moves particles across magnetic surfaces. Then, after the

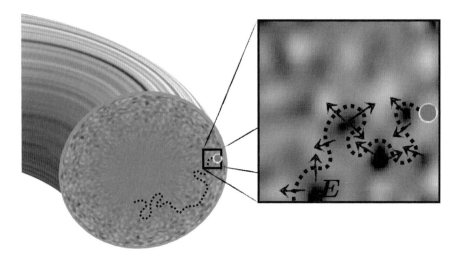

Figure 4.23: The poloidal trajectory of a single ion (dotted line) moving through turbulent eddies (black and white blobs), which are continually popping into and out of existence. The individual turbulent eddies (inset) create short-lived electric fields (solid arrows) that cause particles to undergo an $E \times B$ drift around the eddy. This causes a random walk process, which can transport particles (and the energy they carry) out of the device. Here electric fields point from black to white regions and the magnetic field points out of the page.

---

[29]Proving that the Navier-Stokes equations (which govern turbulence in water) always have a smooth solution is one of the Millennium Prize Problems and will win you a million dollars.

eddy disappears, a new eddy can pop up and move the particle again. This forms the basis for a random walk process, where the particles can jump a distance equal to the size of the eddy during the lifetime of each eddy. Additionally, fluctuations in the flow velocity of electrons can create fluctuating electric currents, which cause small-scale turbulent magnetic fields. This erodes confinement by slightly modifying the externally created magnetic field, leading to minuscule stochastic regions and/or magnetic islands.

Understanding all of this is made more difficult by the fact that turbulence behaves differently along the magnetic field than it does perpendicular to the field. This is because particles can easily travel parallel to the field line, but are confined across it. Moreover, when the plasma is ten times hotter than the Sun, the particles are moving so fast that collisions become infrequent (i.e. the scattering cross-section in Figure 3.9 becomes small). They whiz past each other so quickly that they hardly have time to interact. This means that turbulence can cause fluctuations in the distribution of particle velocities, modifying it from its usual bell curve shape. Hence, to accurately predict behavior, one must go beyond MHD and keep track of the individual particle velocities.[30] Unbelievably, in the last 20 years all of this has been (somewhat) understood using supercomputers and the theoretical model of gyrokinetics, which we will explore in Chapter 6. Unfortunately, this still is not enough. The thing everyone still struggles with is what happens when plasma turbulence comes into contact with a solid material surface. As we'll see, this turns out to be very important for confinement too...

Obviously turbulence poses a challenge, but it also represents an opportunity. We know that there are strategies that can quench turbulence. How do we know this? Because we've stumbled into them before. In 1982, the ASDEX tokamak near Munich, Germany

---

[30] In MHD, collisions are assumed to occur frequently, so the distribution of particle velocities closely follows a bell curve (for a reminder see the "Thermal equilibrium" Tech Box on page 82). That means that only three quantities (the density, the flow velocity, and the temperature) are needed to convey everything about the distribution. In turbulent fusion plasmas the distribution departs somewhat from a bell curve, so we can't use this shortcut.

was performing a routine experiment when it suddenly observed a dramatic reduction in turbulence and a factor of two increase in confinement time. Further investigation revealed that, by taking a few simple operational steps, any tokamak around the world could achieve this improved performance. This mode of operation was dubbed *H-mode* (an abbreviation for "high mode") and it is critical in enabling a future power plant. The physics that underlies it are still fairly unclear and remains the subject of Ph.D. theses to this day.[31]

Because of all the above complications, we don't currently have an accurate way to calculate the confinement time from physical principles. The best we can do is run a whole bunch of experiments at different parameters and measure the confinement time. Then, we plot these experimental results and draw a line that best fits all the data. This produces what is known as an *empirical scaling law*. Effectively, it is a way to extrapolate current experimental results to plasma conditions that have never been tried before. This process involves no physics, only data analysis.

For a typical device with a radius of $r = 1$ meter, the best empirical scaling laws predict the confinement time to be roughly a tenth of a second, a bit shy of our goal. Fortunately, there is substantial wiggle room, both in our estimate and our confinement time goal of 1 second. Furthermore, scaling laws indicate a number of strategies that can increase the confinement time somewhat, including simply making the plasma bigger. However, assuming that scaling laws are exactly true, turbulence makes a fusion power plant appear marginal.

While these scaling laws are verified as much as possible, they can only be tested by existing devices. For that reason, whenever a new device is built there can be considerable uncertainty, especially if it is very different from all existing devices. Hence, it is obvious that empirical scaling laws are unsatisfactory, which is why it is important to learn as much as possible. Turbulence is the enemy and, to defeat it, we must understand it! We will discuss turbulence and H-mode more in subsequent chapters. At this point, it suffices to know that turbulence is what causes our magnetic bottle to leak.

---

[31] Something one of the authors is all too familiar with.

4.14 The Lawson Criterion and the Triple Product

How much confinement is enough confinement? We've stated that, for typical magnetic fusion devices, we need an energy confinement time of roughly 1 second. However, due to turbulence, it looks like a power plant would be marginal. So *exactly* what value of the confinement time should we aim for? Well, there is no single value of the confinement time that is critical to reach. Instead, to make a viable fusion reactor we must bring together a sufficient quantity of three ingredients: density, temperature, and confinement time. In other words, we need to get enough fuel that is hot enough together for a long enough time. To make this precise, we'll turn to work from the 1950s by John Lawson, a physicist working at Culham Laboratory in England.

Lawson developed a metric of success, now known as the *Lawson criterion*, that is independent of the particulars of the confinement scheme and the fusion fuel. He had one clear objective in mind, *a steady-state plasma that can keep itself hot with the fusion energy it produces.* This condition is termed *ignition* because it is analogous to starting a fire. To get a wood fire going, you need to apply some initial energy (e.g. a match), but then it burns for as long as it has fuel. Similarly, once an ignited plasma gets going, it can burn on its own without needing any external heating. It would steadily produce fusion energy for as long as you continue to supply fusion fuel.[32]

TECH BOX: Ignition and the Lawson criterion

Ignition is actually a pretty ambitious goal. This is because all of the fusion energy carried by neutrons immediately escapes confinement. It is only the fusion energy carried by charged particles that stays in the plasma to help keep it

---

(Continued)

---

[32] Lawson's paper can be found in Ref. [19].

(*Continued*)

warm. For example, in D–T fusion, the energy carried by charged particles is only one-fifth of the total fusion energy (i.e. $f_{\text{charged}} = 1/5$). Therefore, to achieve ignition we require that

$$f_{\text{charged}} P_{\text{fusion}} \geq P_{\text{leak}}, \tag{4.2}$$

where P_{leak} is the power that is continually leaking out the plasma and P_{fusion} is the total fusion power that the plasma is producing. Now, P_{leak} can be related to the energy confinement time τ_E through its definition. Additionally, P_{fusion} can be directly calculated from the density n, the temperature T, and the fusion cross-section σ. Incorporating all of this information produces the equation[33]

$$n\tau_E \geq \frac{12T}{\langle \sigma v \rangle} \frac{1}{f_{\text{charged}} E_{\text{fusion}}}, \tag{4.3}$$

known as the *Lawson criterion*. Here E_{fusion} is simple. It is just the total energy released by each fusion reaction (e.g. 17.6 MeV for the D–T reaction). The quantity $\langle \sigma v \rangle$, on the other hand, requires a bit of explanation. It is the fusion cross-section times the particle velocity, then averaged over the velocity distribution of particles (e.g. a bell curve). Remember, in Chapter 3 we learned that the fusion cross-section is an area that indicates the probability of a fusion reaction occurring. When you throw objects with larger cross-sections at one another, they are more likely to strike each other (think of colliding tennis balls versus basketballs). Multiplying the cross-section by the particle velocity is necessary to adjust for the fact that

(*Continued*)

---

[33]Performing this derivation requires the expressions $\tau_E = W/P_{\text{leak}}$ for the confinement time, $W = 3nT$ for the total stored energy in the plasma, and $P_{\text{fusion}} = n^2 \langle \sigma v \rangle E_{\text{fusion}}/4$ for the total fusion power.

> (*Continued*)
>
> fast-moving particles will pass by more particles per unit time. This means they will fuse more quickly than a slow-moving particle that has a fusion cross-section of the same size. An illustration of this fact can be found in *Star Wars*, where Hans Solo and Chewbacca had to do a bunch of calculations to make sure it was safe to jump to light-speed in the Millennium Falcon. When they were cruising around slowly, the chance of colliding with a planet or star was low. However, at light speed, they rush past so many objects so quickly that there was a much higher chance that one would be in their path.

The Lawson criterion says that, to achieve ignition, our device must attain a sufficiently large value of $n\tau_E$ — the plasma density n times the energy confinement time τ_E. All that is necessary to derive this is to set the power leaking out of the plasma equal to the plasma self-heating due to fusion. It is not necessary to specify the confinement scheme or the fuel, so it is very general condition. All ignited plasmas must satisfy the Lawson criterion. This can be done in different ways. For example, the inertial confinement that occurs in hydrogen bombs has a confinement time τ_E of less than a nanosecond, but it makes up for it with enormous densities n. Magnetic fusion, on the other hand, has a relatively modest density, so it requires a much longer confinement time.

But earlier we said that we require three things for a fusion reactor: confinement time, density, and temperature. The Lawson criterion clearly motivates high density and confinement time, but the dependence on temperature is more complex. This is because the value of $n\tau_E$ needed for ignition is different when the plasma is at different temperatures.[34] To illustrate this, Figure 4.24 shows the

---

[34]This can be seen on the right-hand side of Equation (4.3), where temperature appears explicitly, but also enters through the complicated cross-section parameter $\langle \sigma v \rangle$.

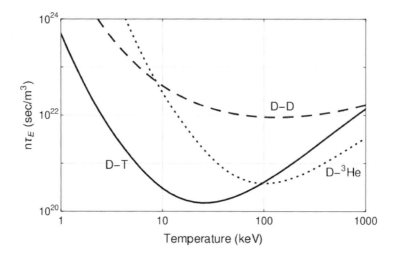

Figure 4.24: The Lawson criterion for three different fusion fuels: D–T (solid), D–D (dashed), and D–^{3}He (dotted). Notice that D–T fusion can be achieved with the lowest value of $n\tau_E$ and at the lowest temperature.

value of $n\tau_E$ required for ignition as a function of temperature for different fusion fuels. We see that it varies strongly, but the lowest possible value is roughly

$$n\tau_E \geq 2 \times 10^{20} \, \text{s/m}^3. \tag{4.4}$$

This minimum occurs at a temperature of $k_B T = 25\,\text{keV}$ for D–T fuel. We will see that, in current magnetic confinement experiments, MHD instability limits the density to around 2×10^{20} particles per cubic meter. Therefore, meeting the Lawson criterion requires an energy confinement time of about $\tau_E = 1$ second (i.e. the goal we set at the beginning of this chapter).

It is possible to simplify the Lawson criterion further, but we lose some generality in doing so. As a happenstance of nature, for D–T plasmas with temperatures between 5 and 20 keV, we can rewrite the Lawson criterion to nicely display the dependence on temperature. We have to make an approximation,[35] but it gives us a beautiful

---

[35] Specifically, we approximate the value of $\langle \sigma v \rangle$ to be proportional to T^2. This isn't exact, but it is accurate to within 10% (for a D–T plasma with a temperature between 5 and 20 keV).

condition known as the *triple product*:

$$n\tau_E T \geq 5 \times 10^{21} \text{ keV-s/m}^3. \tag{4.5}$$

It says that, to make an ignited D–T plasma, we need the product of density, temperature, and confinement time to exceed a single critical value. That's all there is to it.

That said, ignition isn't actually needed to make a commercial power plant, though you do need to get fairly close.[36] Instead, we really care about the parameter $Q = P_\text{fusion}/P_\text{external}$. This parameter is important. It is called the plasma power multiplication factor and it's the amount of fusion power produced by the plasma divided by the amount of external heating power needed to keep the plasma hot. So ignition corresponds to $Q = \infty$ because we don't need any external heating power since the plasma keeps itself hot. However, it looks perfectly acceptable to operate a power plant at say $Q = 30$. This means that you get thirty megawatts of fusion power for every megawatt of external power you supply. This might be preferable because $Q = 30$ operation is slightly easier to achieve, only requiring 85% of the Lawson criterion.[37]

In fusion popularization, a lot of hubbub is made about the condition that $Q = 1$, which is termed *breakeven*. Breakeven indicates that the total fusion power equals the external heating power. While achieving breakeven is significantly easier than ignition, only requiring 15% of the Lawson criterion, it's not a particularly meaningful condition. Nothing special happens when $Q = 1$.[38]

---

[36] Ignition probably isn't even desirable. If the plasma requires some small amount of external heating, it gives the operator a bit more control. Without heating power there is one less knob to turn and so you must rely entirely on other control systems.

[37] The value of $n\tau_E$ required to achieve a certain value of Q is given by $n\tau_E = (n\tau_E)_\text{ignition}/\left(1 + (f_\text{charged}Q)^{-1}\right)$, where $(n\tau_E)_\text{ignition}$ is the value needed for ignition (i.e. the value calculated by Equation (4.3)). This equation can be found by repeating the derivation from the above Tech Box and adding a third term to Equation (4.2) to represent the external heating.

[38] Breakeven isn't even important for a power plant to produce net electricity. In fact, we will see that, if a power plant had perfectly efficient power conversion technology, $Q < 1$ could still lead to net electricity production (see Figure 5.11 and the footnote therein).

Rather, $Q = 5$ is where the action takes place. When $Q \geq 5$, the power carried by the helium from D–T fusion starts to exceed the external heating. The helium is important because, as a charged particle, it is confined by the magnetic field and deposits its energy in the plasma as heating. Thus, when $Q \geq 5$ the plasma is said to be *burning*. Burning plasmas are interesting because the way the plasma heating self-regulates becomes important. Instead of operators directly controlling the heating externally, the plasma takes the reins and begins to heat itself.

4.15 Where is Magnetic Fusion Now?

Take a look for yourself. Figure 4.25 shows the fusion community's progress as measured by the triple product. It's really quite impressive. From 1970 to the year 1997, the triple product doubled every 1.8 years (on average). This outpaced the famous Moore's Law, which observes that the number of transistors on a computer chip doubles every two years. In 1997, the JET tokamak used 24 MW of

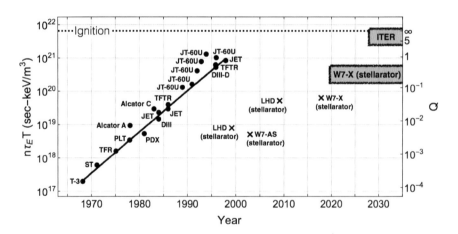

Figure 4.25: The increase in the tokamak fusion triple product (and the equivalent value of Q) was rapid [21], but has stalled since the early 2000s. ITER, the next big fusion experiment, is due to start preliminary operation around 2025 and should achieve considerable improvement (but exactly how much is uncertain). Additionally, several stellarators are shown as well as projections for W7-X, a JET-scale stellarator that began preliminary operation in 2016 [22].

external heating to produce 16 MW of D–T fusion power [20]. This set the world record value of $Q = 0.7$, just shy of breakeven at $Q = 1$.[39]

Unfortunately, progress in the triple product and Q has stalled since then. The causes of this are not entirely clear. One important factor, governmental funding for fusion research, has been steadily decreasing (until recently). However, the most dramatic and obvious change is ITER, the topic of Chapter 7. ITER is an enormous tokamak currently under construction in the south of France. It is the next big thing for fusion. It has been conservatively designed to achieve $Q = 10$ and is the most consequential step the field has ever taken. Since the design was completed in 2001, the worldwide fusion community has been combining resources and preparing technology. Unfortunately, ITER will not add a dot to Figure 4.25 anytime soon. It is not due to start operation until at least 2025.

With ITER in mind, the roll-off in the triple product is made more understandable. The fusion community is at the point where it believes it knows how to achieve a high Q plasma. Unfortunately, doing so requires a huge device that is very expensive, takes a long time to build, and has added organizational challenges. All these issues will be discussed in more depth in Chapter 7. Moreover, because it is such a huge investment, ITER's design must be executed very carefully and with significant margins to ensure success. Finally, its size makes fusion technology, the components that surround and enable the plasma, more challenging than ever before. The advancement of fusion technology is essential to build a power plant, but has only an indirect impact on the triple product. Understanding these technological components will be the focus of the next chapter.

---

[39]The JT-60U tokamak has achieved higher values of the triple product, but not Q. This is because JT-60U doesn't have facilities to handle radioactive material like tritium, so it only ever creates pure deuterium plasmas.

Chapter 5

Fusion Technology

Fusion technology refers to the wide variety of technological components that are needed to enable plasma operation and power extraction. This ranges from the magnets that create the magnetic field, to the heating systems that keep the plasma hot, to the lithium blanket that breeds tritium. Since the tokamak currently has the best performance and is supported by the most developed technology, we will focus on it as an illustrative example. For the most part, the components we will discuss are similar in other fusion devices (especially stellarators[1]).

By now, we have spent quite a few pages (139 to be specific) motivating fusion and the particular geometries and conditions that enable it. Here, we will cut to the chase and describe how all of this could be practically accomplished in a tokamak power plant.

5.1 Magnets

In the previous chapter, we used our knowledge of magnetism to confirm that the magnetic fields in a tokamak are physically possible. To generate the toroidal field, we found that we need a circle of coils, each of which loop around the field (see Figure 5.1). This field, the one that goes the long way around the device, provides the

---

[1]The most significant exception is that stellarators do not need systems to drive plasma current.

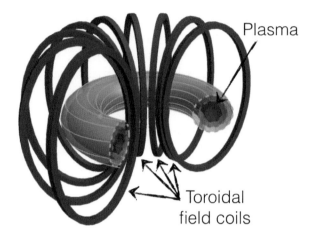

Figure 5.1: The external toroidal field coils that generate the toroidal magnetic field, which provides the dominate force confining the plasma. Why are the coils so much bigger than the plasma? To make space for all the other components we will soon discuss.

dominant force confining the particles, so we want it to be as strong as possible. Therefore, it is unsurprising that these coils need to be beefy. In fact, toroidal field coils are amongst the biggest magnets in the world. Not only can they be several meters in diameter, but they also generate some of the strongest steady-state magnetic fields on Earth. For example, the JET tokamak has 32 toroidal field coils, each of which are over 3 meters in diameter.[2] Together they create a 4 Tesla toroidal field throughout a 100 cubic meter plasma volume. This field is a thousand times stronger than a typical fridge magnet and a bit stronger than the field inside hospital MRI machines.

Creating these magnetic fields poses a challenge. After all, to create tremendous fields, we must circulate a tremendous amount of electric current in the coils. However, most conductors have

---

[2]JET in England and JT-60U in Japan have similar sizes and have been the biggest tokamaks since they were built in 1980s. JET is an acronym for "Joint European Torus," while JT-60U stands for "Japan Torus-60 Upgrade" (though it is in the process of being replaced with JT-60SA or "Japan Torus-60 Super Advanced").

resistivity, meaning that, as electrons move, they bump into the atoms that make up the structure of the coils. This slows the electrons down (thereby reducing the current) and heats the magnets up. To compensate for the resistivity of the material, we must supply power, both to continually top-up the flow of current and to circulate coolant to keep the coil reasonably cool. In most day-to-day applications, this isn't much of a problem, so it is fine to use a run-of-the-mill conductor like copper. However, the insane fields needed for fusion mean that countering resistivity would necessitate an intolerable amount of power. The coils could easily consume hundreds of megawatts of power, a significant fraction of a power plant's electricity output.

For this reason, we turn to magic: superconductivity. Superconducting materials can carry current with exactly zero resistivity. This means that, once we get the electric current flowing in a coil, it will circulate forever without needing any external help. Moreover, the presence of the current doesn't cause the coil to heat up. While the physical principles underlying this are fairly well understood (which we will discuss in the next chapter), superconductivity is about the closest thing our world has to magic. It enables a dramatic reduction in the amount of power needed to create our magnetic fields. In doing so, we have to make a few sacrifices, but they are usually relatively minor. For example, superconducting material is more expensive and fragile than conventional conductors. Additionally, a material only becomes superconducting when its temperature is below a certain critical value. This value depends on the material, but is very low, typically just tens of degrees above absolute zero.[3] This leads to what is undoubtedly the largest steady-state temperature gradient in the Solar System (and maybe the universe): 50 million°C in the center of the plasma and $-269°C$ in the coils just a meter away.[4]

---

[3] Absolute zero is the lowest temperature possible. It corresponds to $-273.2°C$, $-459.7°F$, or 0 Kelvin. At absolute zero, all particles are as close to stationary as quantum mechanics allows them to be.
[4] In 2003, the Tore Supra tokamak (now WEST) in France achieved these conditions for over 6 minutes.

Nevertheless, making the coils of a fusion device superconducting saves hundreds of megawatts of power, making these difficulties well worth the headache.

> **TECH BOX: Why superconductivity is basically magic**
>
> Mother Nature rarely gives us exact solutions to our human problems. Normally, we have to painstakingly optimize a system in order to bring what is physically allowable as close as possible to what we desire. Now, imagine the perspective of an electrical engineer first learning about superconductivity. They've spent their entire professional lives designing circuits to minimize power consumption by incrementally improving things like conducting materials, the geometry of wiring, efficient cooling techniques, etc. Then, along comes a non-intuitive phenomena that just so happens to exactly solve all these problems. Someone shows up with a special ceramic wire and says "as long as you keep this wire below $-254.8°C$, it will conduct electricity without any losses or consuming any power." Superconductivity is one of the few instances in human history when an engineering problem was found to have an exact physics solution. The other example that comes to mind is the compass, which gave seafarers exactly what they needed: a fixed directional reference anywhere on the oceans. We say these things are magic, not because they are inexplicable, but because they elegantly and completely solve a human problem. Typically, the way the world works only partially overlaps with the way *we wish* the world worked.

For the above reasons, nearly all proposed tokamak power plant designs use superconducting coils, as do several current experiments. However, coil design involves other challenges apart from electrical resistivity. One of the most important is the mechanical stress experienced by the coil. As the flow of current is increased, opposite sides of the coils push on each other via magnetic interactions, which can make the coil blow itself apart.[5] To counter this, the coil must be designed to minimize these forces and includes large amounts of structural steel to buttress against them.

While the toroidal field coils are responsible for generating the dominant magnetic field, they aren't the only coils in a tokamak. If you look at an actual tokamak, you will notice poloidal field

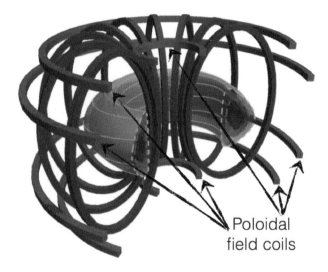

Figure 5.2: The external poloidal field coils, which generate poloidal magnetic fields, can modify the cross-sectional shape of the plasma. Here, we show the plasma with the common "D"-shaped magnetic surfaces, as opposed to the circular magnetic surfaces we saw in Figure 5.1.

---

[5]The force that causes a current-carrying loop of wire to tend to blow itself apart is called the *hoop force*. It can be understood by remembering that the magnetism pushes particles in the direction perpendicular to both the particle's direction of motion and the magnetic field. We know that in a circular magnet, the electrons are moving azimuthally along the magnet, while the magnetic field points axially through the wire loop's center. Hence, the magnetic force pushes the electrons (and the wire as a whole) in the radial direction, specifically outwards.

coils (shown in Figure 5.2). These poloidal field coils run current in the same direction as the plasma current and can be used to modify the shape of the plasma.[6] If you don't use any poloidal field coils, the plasma will be shaped like a normal donut. This means that, if you put this donut down on a table and cut a wedge out, you will find that it has a roughly circular cross-section (as is shown in Figure 5.1). By using the poloidal field coils to apply an external poloidal magnetic field, we can change this cross-sectional shape of the magnetic surfaces. As we will discuss in Chapter 6, some magnetic surface shapes (e.g. the "D"-shape shown in Figure 5.2) have been found to substantially improve plasma performance. Note that, regardless of what we do with the poloidal field coils, the plasma necessarily remains symmetric in the toroidal direction.

There is one last major magnet system. But, instead of being used to directly generate a steady-state magnetic field, it is used to drive the current in the plasma and provide heating.

5.2 Plasma Heating and Current Drive

In the last chapter, we learned that, to defeat particle drifts, we must create a poloidal magnetic field (i.e. a field that wraps around the device in the short direction). In a tokamak, this cannot be accomplished with the external poloidal field coils alone — you must use the plasma itself as a poloidal field coil by finding a way to drive an electric current in it. How is this done? It's tricky!

There are several methods that are often used, all of which also heat the plasma. Remember, in addition to driving current, we must increase the temperature of a bunch of D–T gas to hundreds of millions of degrees. That requires a lot of heating. The following

---

[6]Importantly, the poloidal field coils are also necessary to ensure the stability of the plasma. Like solid magnets, the current running through the plasma also experiences the hoop force, which pushes it in the outboard direction. The poloidal field coils are needed to create a uniform vertical field across the entire tokamak, which counters the hoop force and keeps the plasma in equilibrium.

three techniques are the most common and each have their own set of advantages and disadvantages:

- inductive,
- neutral beam, and
- electromagnetic wave.

5.2.1 *Inductive*

The most commonly used component for heating and current drive is shown in Figure 5.3: the *central solenoid*. In the center of almost all tokamaks, there is a long straight vertical line of circular coils. By continuously increasing the electric current running through these coils, we can use electromagnetic induction to drive the plasma current (the details of this are given in the following Tech Box). Not only does this create the poloidal field needed to neutralize particle drifts, but it also provides heating. This works because, while plasma is a good electrical conductor, it's *not* superconducting. Therefore, the electrons that carry the current collide with each other and the ions as they move. These collisions give the plasma resistivity and cause electrical energy to be converted into thermal energy.[7] Early tokamaks relied exclusively on the central solenoid for both driving current and heating.

While the specifics of inductive heating and current drive are complex, the most important detail is that it requires us to *continuously increase* the current running through the central solenoid. This is a problem because it cannot be done indefinitely. No matter how big our central solenoid coils are, we will eventually reach a limit (e.g. the coils blowing themselves apart) and will be forced to stop. As soon as we stop driving current, the particle drifts jump into

---

[7]Note that inductive heating becomes less effective in hot, high-performance fusion devices (e.g. a power plant). This is because the scattering collision cross-section becomes smaller when the particles are moving faster (as shown in Figure 3.9 in Chapter 3).

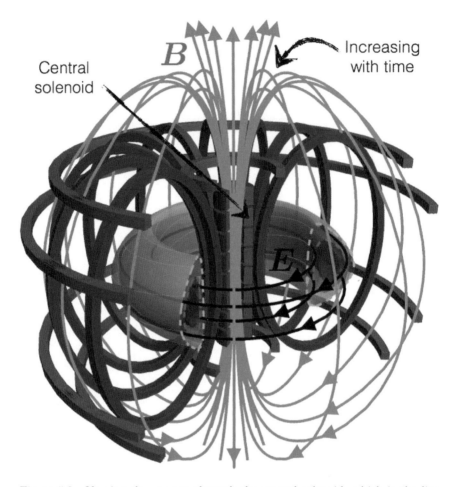

Figure 5.3: Varying the current through the central solenoid, which is the line of coils running through the donut hole of the device, drives a toroidal current in the plasma. The magnetic field (gray lines) and the induced electric field (black lines) generated by the central solenoid are also shown.

action and destroy our plasma confinement. Hence, the important takeaway message is that, while inductive current drive was very useful for the pulsed experiments in the early days of fusion, its future prospects aren't as good. While it is efficient and reliable, it can only drive current for a short time (i.e. a few minutes at most). Since we would really like a fusion power plant to be steady state, designs typically only include a small central solenoid to help jump-start plasma discharges.[8]

TECH BOX: Inductive heating and current drive

As its name suggests, inductive heating and current drive fundamentally rely on the electromagnetic induction that we learned about in Chapter 1 and on page 96. Remember that a time-changing magnetic field induces an electric field that circles around it. However, for heating and current drive, induction is just one part of the overall process:

(1) We build a line of circular coils running up through the donut hole of the tokamak.
(2) During operation, we continuously increase the electric current in them.
(3) This creates a vertical magnetic field that strengthens with time.
(4) The time-changing magnetic field induces a toroidal electric field.
(5) This electric field pushes the electrons in the plasma, creating a plasma current.
(6) The plasma current generates the poloidal magnetic field and provides heating due to the plasma's resistivity.
(7) The poloidal field neutralizes the particle drifts that would otherwise destroy our plasma.

Complicated much? Yes.

---

[8] You could imagine a fusion reactor continually repeating long pulses in order to get by with inductive current drive. However, pulsed operation is harder on the components around the plasma (due to material fatigue from stress and thermal cycling) and has worse economics (since power production is only occurring part of the time).

5.2.2 Neutral beam

Beam heating, or shooting a beam of really fast particles into the plasma, is perhaps the most obvious way to provide heating. It is similar to blowing hot air into a room to heat it up and is exactly what we considered when discussing thermal equilibrium in Chapter 3. The idea of beam heating is to set up an accelerator next to our tokamak and shoot in a beam of particles, each with more energy than the average particle in the plasma. As these beam particles collide with the plasma, they slow down and share their energy. Not only does this cause heating, but it also refuels the plasma and can drive current.

The tricky part is that particle accelerators typically work by using an electric field to accelerate charged particles. Charged particles, however, would be deflected by the strong magnetic fields of a tokamak. So, if we try to use a normal particle accelerator, the particle beam will be deflected before it can even get to the plasma. To deal with this, fusioneers (i.e. someone who attempts nuclear fusion) must build a special type of particle accelerator called a neutral beam injector. These are like normal accelerators, except there is an extra step at the end: neutralizing the particle beam. To do this, the beam of ions from the accelerator is passed through a gas of neutral atoms where it picks up the electrons it needs for neutrality. Afterwards, any residual charged particles are filtered out using a weak magnetic field that slightly deflects them out of the straight path of the neutral particles. Finally, the entirely neutral beam is sent to the plasma (see Figure 5.4).

An important aspect of the neutral beam is choosing the speed the individual particles are accelerated to. This determines how far into the plasma the particles will make it before they ionize due to scattering collisions. As is intuitive, a beam made of faster particles will penetrate further (or the same distance into a more dense plasma). If the beam is too fast it can pass entirely through the plasma, while, if it is too slow, the beam will be stopped near the edge. By deliberately choosing the beam speed, we can ensure that particles and energy are deposited primarily where we want them (i.e. near the center of the plasma).

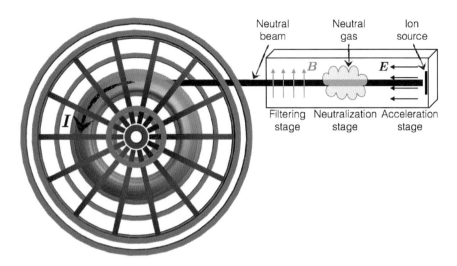

Figure 5.4: A top view of a neutral beam injector (on the right) heating and fueling a tokamak (on the left), as well as driving a plasma current I. Yes, neutral beam injectors are often almost as big as the tokamak itself.

As soon as the beam ionizes in the plasma, it becomes a directed flow of ions. Since ions are charged, they embody an electric current. If we angle the neutral beam injector to shoot toroidally into the device (as is shown in Figure 5.4), we can create plasma current. Importantly, this process can be sustained in steady state because it does not rely on electromagnetic induction.

While the inner workings of a neutral beam injector are simple to state, in practice, building one that operates efficiently is a challenging technological endeavor. At each step, there are losses. Forming a beam of ions in the first place is inefficient, this beam can't be accelerated perfectly, and the neutralization process is never 100% effective. Neutral beam injectors on existing experiments only deliver about 30% of the power they draw from the electrical socket to the plasma [36]. Nevertheless, development is ongoing, both to improve their efficiency and to achieve the high-speed beams needed for the large, dense plasmas of a future power plant. While neutral beams are the dominant method of *non-inductive* current drive in existing experiments, it is unclear if their popularity will continue.

5.2.3 *Electromagnetic wave*

The final way to drive current and heat a plasma is with electromagnetic waves. This is the same principle that a microwave uses to reheat your leftovers. An electromagnetic wave is a self-sustaining oscillation of electric and magnetic fields that can even propagate in the vacuum of empty space (see Figure 5.5). The magnetic field (which is changing with time) creates an electric field through induction. Then, the electric field (which is also changing with time) gives rise to a magnetic field through an analogous mechanism. This process repeats, moving the wave forwards in space. All the colors of the rainbow are just electromagnetic waves with different wavelengths (red has the longest length and violet the shortest). Moreover, there are electromagnetic waves with wavelengths outside of what the human eye can see, many of which are useful for fusion devices.

Imagine launching an electromagnetic wave into a plasma. We learned that, due to quasineutrality, the individual electrons and ions in the plasma respond to cancel out electric fields. However, this process is neither exact nor instantaneous (see the Tech Box on page 98 for a reminder). So, if the electromagnetic wave oscillates

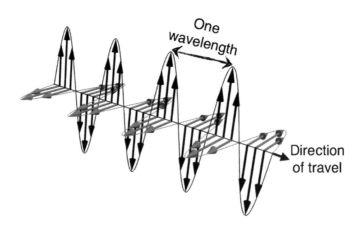

Figure 5.5: The electric (black) and magnetic (gray) fields of an electromagnetic wave. These electric and magnetic fields oscillate up and down or side to side in time, respectively.

extremely quickly and has a very short wavelength, the particles in the plasma will not have time to respond and the wave will propagate just like it does in empty space. However, if the oscillation time of the wave is comparable to oscillations of particles in the plasma (e.g. the time it takes a particle to oscillate around a magnetic field line), then interesting stuff can happen. Particles in the vicinity of the electromagnetic wave will feel its fields and respond with their own electric and magnetic fields, which can modify the electromagnetic wave. This can have several possible outcomes. The plasma can reflect the wave like a mirror, the plasma can change the wave's characteristics (e.g. wavelength, oscillation time, speed, etc.), or the plasma can even absorb the wave. We are most interested in this last case because it provides plasma heating. However, to successfully employ electromagnetic waves, all these effects must be understood and taken into account.

In order for plasma to absorb a wave, there must be a net transfer of energy from the wave to the particles. Usually, this doesn't happen because the electric field of the wave oscillates back and forth in opposing directions. Hence, particles are accelerated one way, but then are accelerated back the other. On average, no net acceleration occurs. However, in certain circumstances, this isn't true. For example, Lev Landau, a colorful Russian theoretical physicist, showed that under certain conditions, particles moving along with a wave won't experience its oscillation. Instead, the particle can experience a constant electric field, which will provide net acceleration. This is called *Landau damping* and a common analogy is that of a surfer and ocean waves. If a surfer is sitting stationary out in the ocean, she gets pushed back and forth by the waves and nothing too exciting happens. However, if the surfer is moving with the right velocity and times things well, she can surf a wave crest, gaining energy from the wave until she reaches the shoreline. Similarly, if a plasma wave has the right speed, it can transfer its energy to particles in the plasma.

Another way to accelerate particles has to do with matching oscillations in the plasma with the oscillations in the electromagnetic wave. For example, we know that particles in a tokamak oscillate around magnetic field lines. If the electromagnetic wave oscillation matches this particle oscillation, then, even though the wave is pushing back and forth, it will always be pushing in the direction that the particle is moving. Hence, net acceleration! This works a lot like swinging on a swing set. You kick your legs forwards and backwards, but, as long as each kick is timed to reinforce your direction of motion, you can make yourself move faster and swing higher. Examples of this include *electron cyclotron heating* and *ion cyclotron heating*, which match the gyration frequency of the electrons and ions, respectively. Generally, electromagnetic wave heating can deliver about 40% of the power they draw from the grid to the plasma — a somewhat higher efficiency than neutral beam injectors.

Additionally, if we are even more clever, waves can be used to drive an electric current. Remember, current is just the net motion of charged particles. The faster the particles move, the higher the current. If two identical charged particles are moving in opposite directions, there is no current (since on average, the particles are stationary). But if we can use waves to accelerate just one of them, then a current arises (see Figure 5.6). Hence, by tuning our waves such that they preferentially match the oscillations of electrons traveling in one direction along the magnetic field line, we can drive the plasma current. This is called *electron cyclotron current drive* and it is one of the several possible methods.[9]

Lastly, we can control where heating and current drive occurs in the tokamak by tuning the waves such that they are absorbed only for particular plasma conditions (i.e. at certain values of the magnetic field, plasma density, and plasma temperature). This requires a lot of analysis though because the waves are launched by metal antennae near the edge of the plasma. They then must propagate from a region with no plasma to the point where they get absorbed. All the while, we must ensure that the waves don't get reflected, accounting for the fact that they are being modified by the plasma as they travel.

---

[9]For example, we can also drive current using Landau damping (i.e. the surfing example).

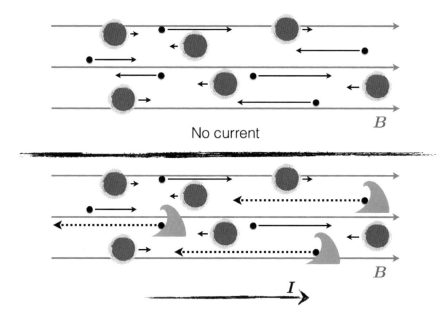

Figure 5.6: When particles have no preferred direction the current is zero (top), but, by accelerating the electrons traveling in a particular direction (dotted arrows) with an electromagnetic wave, we can drive current (bottom).

Though heating and current drive with electromagnetic waves may seem hopelessly complex, it is commonly employed in existing devices. Still, there remains much room for progress, both in improving the technology that creates the electromagnetic waves and in understanding the way they interact with the plasma.

5.3 First Wall

The first wall is the solid material surface that directly faces the plasma (see Figure 5.7). Because of its location, it is the primary component that must be built to withstand unexpected plasma excursions. In a tokamak, the most extreme scenario is called a *disruption*, which is an uncontrolled and abrupt loss of confinement. In current experiments, the entire plasma typically loses MHD stability and moves until it touches the nearest solid object — the

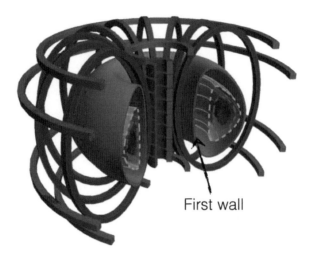

Figure 5.7: The tokamak first wall, which is the closest solid component that surrounds the plasma.

first wall. Disruptions are driven by the plasma current[10] and can be triggered by a number of mechanisms. Rare events like a bit of metal falling into the plasma or the malfunction of a magnet almost always result in a disruption. More commonly, if we accidentally try to confine too much fuel or drive too much current, the plasma goes MHD unstable and disrupts (we will learn more about these stability limits in Chapter 8).

While disruptions are not too worrisome in existing experiments, they become much stronger in larger machines. In a power plant, they will be one of the most serious concerns because they have the potential to damage parts of the device. Though the plasma has a very low density, this destructive power comes from the extreme temperature and large plasma current. Even after the fusion reactions stop, the plasma's thermal and electrical energy can still do substantial damage. The effect of the thermal energy is straightforward — it can erode and/or melt parts of the first wall. The effect of the

---

[10]Stellarators, which lack a plasma current, do not observe disruptions. For this reason, the first wall in a stellarator does not need to be quite as robust as in a tokamak.

electrical energy is more subtle, but just as harmful. As anyone who has stuck a fork into an electrical socket knows, electricity will follow the path of least resistance. After the plasma comes into contact with the wall, it cools down and its resistivity increases. Eventually, the plasma becomes more resistive than the first wall. When this happens, the plasma current migrates into the wall and runs amok.[11] It interacts with the background magnetic field created by the external coils, leading to electromagnetic forces that can break pieces off and crumple the antennae used for plasma heating.[12] In the JET tokamak, these forces once caused the whole device to jump a few millimeters into the air [23]! While this has prompted strategies to reduce the occurrence and severity of disruptions (see the following Tech Box), it also motivates a robust first wall.

TECH BOX: Mitigating disruptions

In existing tokamaks, a disruption is a manageable curiosity, but, in a power plant, it would be a fairly devastating event. Disruptions will need to be almost entirely avoided, but it currently looks like some will probably still occur. When they do, they must be heavily mitigated. The importance of this task for a power plant is not lost on the fusion community. The first step in mitigating a disruption is to design computer systems that can reliably predict their occurrence as early as possible. To this end, scientists have identified a number of so-called "precursor events" that often signal the approaching doom of a disruption. These include things like specific, small

(Continued)

---

[11] Strictly speaking, even before the plasma touches the wall, its motion can create electric currents in the wall via induction.
[12] Additionally, though this is not at all obvious, the rapidly changing plasma current can act like a particle accelerator and create energetic beams of electrons. These are called *runaway electrons* and they can burn small holes through the surrounding components.

(*Continued*)

oscillations in the plasma position and the appearance of magnetic islands.

Ideally, once an impending disruption has been detected, you would gracefully change your operating parameters in order to avoid it. However, there is rarely enough time, typically only milliseconds. Instead, operators rely on automated systems that attempt to minimize the negative impact of the disruption. Typically, this involves trying to dissipate the plasma's energy as quickly as possible. The most promising method, which is called *shattered pellet injection*, is exactly what it sounds like. Pellet guns are stationed just outside of the first wall and can fire large frozen pellets made of deuterium, argon, neon, or helium into the plasma. Just before exiting the pellet gun, the pellet is shattered into shards. This system is optimal because the solid pellets can be fired at high velocities, while the shards have very high surface area such that they dissolve into the plasma quickly. Once in the plasma, the large amount of neutral atoms from the pellet will stop the plasma current and radiate away the thermal energy, spreading it evenly over the whole first wall. While shattered pellet injection appears promising, disruptions remain a big challenge due to their unpredictability and destructive potential.

Lastly, in addition to the disruption forces, the first wall must also tolerate an incredible neutron flux. This is because the wall stands between the birthplace of the fusion neutrons (i.e. the plasma) and where we want them to go (i.e. the tritium breeding blanket). Hence, the fusion neutrons necessarily travel through the wall. While most stream through with little interaction, a lot will collide with atoms in the solid material along the way. In fact, it is estimated that, in a power plant, each atom in the wall would be struck by a neutron a dozen times per year. While this may not sound like much, each collision displaces the atom to a new location in the lattice structure of the solid. Hence, this process gradually rearranges all the atoms of the wall, which can have severe consequences for its mechanical strength. This is troubling because mechanical strength of the material is very important. The wall must be able to withstand disruptions, but we can't make it too thick. If the reactor wall is thicker than a few centimeters, it will block too many neutrons from reaching the blanket, preventing you from breeding enough tritium to fuel the reactor. Unfortunately, the details of this mechanical degradation are still fairly unknown because we don't yet have the ability to experimentally test materials in a fusion neutron environment. The best we can do is to use neutrons from *fission* power plants and assume the materials will respond similarly. Accordingly, first wall designs generally employ special types of radiation-resistant steel that were developed for fission reactors.

5.4 Divertors

A divertor is a specially-designed, replaceable system on the first wall that handles the constant run-off of particles and energy from the plasma. We know that, because of turbulence, our magnetic bottle is inherently leaky. This leakiness is substantial, so we must be careful to avoid things melting and prepared to replace components when damage occurs.

To understand the divertor, we must first understand what determines the edge of the plasma. In the early days of fusion, the edge was defined by a solid block of metal called a limiter, which directly touched the plasma (see the left side of Figure 5.8). As turbulence transported particles outwards from one magnetic surface to another, they would eventually reach a magnetic surface that intercepted the limiter. Remember, in Chapter 4 we learned that magnetic surfaces are the nested surfaces that are mapped out by magnetic field lines. This means that a particle can travel within their magnetic surface very quickly because they can do so by moving along the field line. So as soon as turbulence moves a particle to a magnetic surface that intercepts the limiter, the particle would immediately hit it and deposit its energy. This formed a natural plasma boundary — magnetic surfaces that do not intercept the

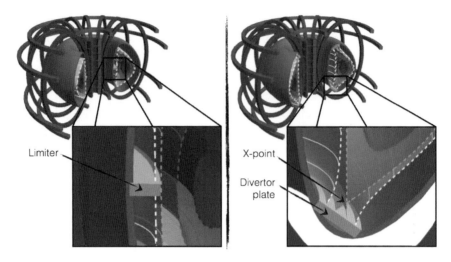

Figure 5.8: Two methods of defining the edge of the plasma: a limiter (left) and a divertor (right). A limiter determines which magnetic surfaces are open (shown in white) with a solid piece of metal, while a divertor uses a magnetic X-point.

limiter (referred to as *closed* magnetic surfaces) can confine plasma and sustain fusion, while magnetic surfaces that touch the limiter at any point (referred to as *open* magnetic surfaces) cannot. This configuration, however, is not ideal because it places solid material up against the edge of the plasma. Intuitively, this seems like trouble because you want the edge of the plasma to be as hot as possible, but the limiter is right next door and you can't let it melt. This intuition is right. What happens is that particles from the plasma crash into the limiter and kick out low-temperature neutral atoms, which then spray into the plasma edge and cool it down.

As the science of fusion matured, it was discovered that better plasma performance could be achieved by defining the edge using a magnetic X-point (see the right side of Figure 5.8). An X-point is just what it sounds like: a magnetic surface intersects itself, making the shape of the letter "X" in the poloidal plane. This change in the topology of the magnetic surfaces can be created by external poloidal field coils and provides a natural division between open and closed magnetic surfaces. The surfaces within the one with the X-point confine plasma, while the ones outside do not. Hence, when turbulence makes a particle jump across the last closed magnetic surface, it quickly streams along the magnetic field line and hits a solid metal divertor plate. This geometry is difficult to describe in words, but please have a look at the right side of Figure 5.8. By keeping the X-point fixed while moving the divertor plate, we see that the solid material can be placed as far from the hot plasma as we wish. Moreover, the divertor plate can be angled such that the cool neutral particles that get knocked out of the solid material have trouble getting into the plasma.

Advances in plasma exhaust have been instrumental in improving the performance of fusion devices. By further separating the plasma from solid material, high plasma temperatures can be more easily achieved. However, we may fall victim to our own success. We want lots of fusion power, but then we are required to manage it as it comes out of the plasma onto the divertor plates. To estimate the scale of the exhaust problem, we must know the amount of heat coming out of the plasma and the area over which it is deposited. The amount of heat exhausted from the plasma is a direct consequence of how much fusion power the device produces. Remember, a fifth of all fusion energy is carried by charged particles (rather than neutrons), which will be confined by the magnetic field and eventually pass through the divertor. Even in the smallest fusion power plants envisaged, this is at least 100 MW. Now looking at Figure 5.8, we see that all of this energy is deposited in a thin area. Because the magnetic surfaces are super effective at confinement, particles move to new magnetic surfaces much more slowly than they travel within them. This means that, as soon as a particle reaches the very first open magnetic surface, it will hit the divertor plate and transfer its energy. Hence, all 100 MW leaving the plasma will be concentrated around a thin toroidal line where the very first open magnetic surface intersects the divertor plate.[13] In present-day experiments, this line can be as

---

[13] Ideally, we would like particles to travel outwards across *closed* magnetic surfaces slowly, but then travel outwards across *open* magnetic surfaces rapidly. That way plasma exhaust would be spread over a thicker toroidal line. Unfortunately, it is difficult to achieve these two goals simultaneously.

thin as a few millimeters! Since the toroidal circumference of the divertor plate is $2\pi R$ (where R is typically envisaged to be around 5 m), the total area is roughly 0.1 m^2.

In summary, we have 100 MW being deposited across two divertor plates, each with a deposition area of just 0.1 square meters. This suggests a heat flux of 500 MW per square meter — an absolutely enormous value! Fortunately, there are some geometrical tricks that we will discuss in Chapter 7 that can give us a factor of about 50 reduction. Nevertheless, 10 MW per square meter is still at the limit of what is possible. It is actually larger than the heat flux experienced by space capsules as they reenter Earth's atmosphere. For this reason, divertor plates are usually made from tungsten (the element with the highest melting point) and have sophisticated cooling systems that pump coolant at high velocities through channels that honeycomb the solid material. In the future, tokamaks may use more advanced divertor designs employing liquid metal or even a thick layer of neutral gas that shields the solid material.

5.5 Tritium Breeding Blanket

The tritium breeding blanket sits directly behind the first wall (see Figure 5.9) and is responsible for absorbing the neutrons generated

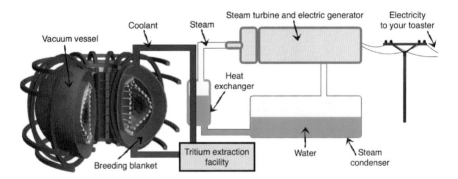

Figure 5.9: The tritium breeding blanket captures the fusion neutrons, which carry most of the energy from the reactor. This energy (along with the energy exhausted to the divertor) is transferred to a coolant, which is used to boil water and run an electric generator.

by fusion in the plasma. In doing so, it must accomplish three tasks: breed tritium to use as fuel, capture the energy of the neutrons to generate electricity, and protect the fragile components behind it from the neutrons.

Tritium is radioactive has a half-life of 12 years, meaning that there is little of it in nature. So, if we want our fusion reactor to run on a deuterium–tritium fusion fuel cycle, we must find a way to make tritium. To do this, power plant designs include a thick blanket containing lithium that surrounds the plasma. Then, the neutrons produced by D–T fusion will travel into the blanket and produce tritium via the lithium breeding reaction:

$$^6\text{Li} + {}^1\text{n} \to {}^4\text{He} + {}^3\text{H} + 4.8\,\text{MeV}. \tag{5.1}$$

This reaction is optimal because it has the largest tritium breeding cross-section of any naturally-occurring isotope.[14] Theoretically, everything works very elegantly. Each D–T fusion produces one neutron, which we can then use to induce a breeding reaction to make one tritium nucleus and some extra energy. Note that the energy produced in breeding, 4.8 MeV, is smaller than the 17.6 MeV produced by D–T fusion, but is still a substantial bonus.

In practice, however, producing as much tritium as is consumed is a challenge. There are all sorts of neutron loss mechanisms. First, the neutrons have to travel through the first wall, which is typically made of metal. Invariably, atoms in the wall will capture some of the neutrons as they pass by, preventing them from breeding tritium. Additionally, the blanket never completely surrounds 100% of the plasma. There are ports in the blanket for heating/refueling systems and the blanket itself usually has structural components. All of these things absorb neutrons or let them stream out of the reactor.

---

[14]Natural lithium is only about 8% lithium-6, while the rest is lithium-7. Lithium-7 can also breed tritium, but each reaction consumes 2.5 MeV of energy and has a significantly smaller cross-section. Hence, for fusion we would like to separate the two isotopes and use mostly lithium-6. Coincidentally, the fission industry already performs this separation because it wants to use only lithium-7 in order to *minimize* tritium production.

Together, these effects can easily pilfer away 30% or more of our hard-earned neutrons. Even after carefully optimizing a design to minimize losses, we still come up short. For this reason, to achieve tritium self-sufficiency, we must include materials that are capable of multiplying neutrons.

Neutron multipliers (as they are called) undergo reactions that produce more neutrons than are required to induce them. The best neutron multiplier for fusion is the beryllium reaction[15]:

$$^9\text{Be} + {}^1\text{n} \rightarrow 2\ {}^4\text{He} + 2\ {}^1\text{n} - 1.9\,\text{MeV}. \tag{5.2}$$

We see that a single neutron breaks apart the beryllium and produces two neutrons. This is exactly what we need, though the process does cost a bit of energy (which must be supplied by the incoming neutron). Once again, in theory, this all appears very elegant, but in practice it's a bit of a mess. Other reactions can occur when a neutron hits beryllium (e.g. neutron capture), which mean you can't count on beryllium to multiply neutrons by too much. Even achieving a neutron multiplication factor of 1.5 is difficult. However, a workable solution does appear possible — each tritium nucleus that fuses in the plasma would lead to at least one tritium being produced in the blanket.

In capturing the neutrons, the blanket also receives their energy, which is its second purpose. This is 80% of the fusion power (remember the remaining 20% stays in the plasma and is handled by the divertor). Fortunately, unlike the divertor, this energy is deposited over a large volume, so it isn't as difficult to manage. Regardless, the blanket must include space for coolant in order to carry large amounts of energy out of the reactor. This energy (along with the energy from the divertor) can be used to boil water, run a steam turbine, and generate electricity.

---

[15] A potential alternative is lead, which actually has a higher neutron multiplication cross-section than beryllium. However, the reaction consumes more than 7 MeV of energy and produces the radioactive isotope polonium-210. You don't want polonium-210.

The final responsibility of the blanket, protecting the components behind it from neutrons, seems like it should be guaranteed by achieving our first two tasks. However, superconducting coils are especially sensitive to neutrons. When neutrons knock into the atoms of the superconductor, they disrupt the precise crystalline structure, gradually reducing the amount of electric current that can be carried. Additionally, since superconductors must be kept extremely cold, any heat deposited in them by the neutrons must be removed. This can be problematic because superconductors operate at near absolute zero and the colder an object is, the more electricity it takes to remove heat.[16] These considerations put a fundamental limitation on how thin a blanket can be — half a meter at bare a minimum. Any thinner than this and too many neutrons will make it through.

While fusion scientists are confident that a satisfactory tritium breeding blanket can be built, it is probably the least developed component in a fusion reactor. Unlike magnets, particle accelerators, and electromagnetic antennae, the tritium breeding blanket currently has no application outside of fusion systems. Moreover, to test one, the plasma must be producing a lot of neutrons for an extended period of time. Because of these constraints, no blanket design has ever been tested in an integrated fashion. So, for the most part, the particulars of blanket design are still up in the air. One design uses a tank of many small lithium pebbles, intermixed with neutron-multiplying beryllium pebbles. After a tritium atom is created in the lithium pebble, it is no longer bound by the lattice structure of the solid material. The tritium naturally migrates through the solid in a random walk process and, because of the small size of the pebbles, quickly escapes. All the while, helium gas is continually pumped through the pebble tank, in order to whisk away the newly formed tritium and serve as a coolant. After leaving the blanket, the gas is used to boil water and the tritium is separated out to use for

---

[16]To understand why is left as an exercise for the reader. Hint: check out the "Thermodynamic efficiency" Tech Box on page 31 and remember that electricity is just the directed motion of electrons.

refueling. Another blanket design clads the outside of the first wall in solid beryllium and immerses the whole thing in a big tank of liquid containing lithium. In this design, the lithium compound serves as both the coolant and the breeding material. This is attractive because it minimizes the amount of neutrons captured by solid structural material. While all these widely varying designs appear to work in theory, what we really need to distinguish them is experimental testing.

5.6 Vacuum Vessel

Just outside the blanket sits the vacuum vessel (see Figure 5.9). It's a big metal can that prevents air from getting in, which illustrates just how fragile and contingent the plasma is. When people think of nuclear power, their intuition is of an unstoppable force, a force that scientists struggle to prevent from running wild. While this may have some merit for *fission* power, it couldn't be further from the truth of fusion power. Fusion power is a lot like me when I was 12 years old: delicate, shy, and picky.

In order to create the plasma, we must build a vacuum vessel and pump all of the air out of it. Air is about a million times more dense than a typical magnetic confinement fusion plasma, so, if we didn't expel it, it would contaminate things and prevent our heatings systems from achieving fusion-relevant temperatures. If our vacuum vessel sprang a leak during operation, the massive influx of air (relative to the amount of plasma) would immediately cool everything down and stop all fusion reactions. It would be like dumping a tub of water on a lit match.

5.7 Diagnostics

In fusion *experiments*, diagnostics are king. Diagnostics are our window into the world of the plasma. In practice, this window is heavily frosted, requiring sophisticated technology and careful interpretation to establish even basic observations. After all, fusion plasmas are ten times the temperature of the Sun. You can't exactly put your hand to its forehead to see how warm it is.

Modern diagnostics allow us to precisely measure the plasma's performance, compare theoretical ideas against reality, and even implement real-time feedback (i.e. use the current behavior of the plasma to instantaneously adjust our plasma control systems). However, in the early days of fusion, diagnostics were much more limited. Small loops of wire were placed outside the plasma to measure the magnetic field. This would alert experimenters to the termination of the plasma because the poloidal field generated by the plasma current would vanish. Small metal rods called *Langmuir probes* were stuck into the plasma edge, yielding information about the density, temperature, and electric field there.[17] Unfortunately, Langmuir probes were not able to survive the high temperatures near the center of fusion plasmas and their very presence could alter the behavior they were measuring. Finally, neutron detectors, which measure the number of neutrons coming out of the plasma, gave an overall indication of how many fusion reactions were occurring.

While these diagnostics were better than nothing, the limited information they provided could certainly lead to misinterpretation. The most infamous example of this was the ZETA experiment, a device similar to a tokamak that was built in England in 1957.[18] Soon after experiments started, the neutron detectors in ZETA began to measure lots of fusion neutrons from the millisecond long plasma discharges. This seemed to imply high plasma temperatures and unprecedented performance. After a few months of analysis and review, the results were published and a press conference was held, which caused a worldwide sensation. Headlines ran. Politicians took credit. But ultimately, it was all for naught. The results turned out

---

[17] Irving Langmuir, inventor of the Langmuir probe, was a Nobel laureate and one of the fathers of plasma physics. In fact, he is responsible for the name "plasma." While studying fluorescent light bulbs in 1927, he observed this new state of matter ... and it reminded him of blood plasma! Why was never entirely clear, even to his close collaborators.

[18] The ZETA experiment was what is now called a reversed field pinch. Like a tokamak, they have a donut shape and a plasma current, but their toroidal magnetic field is much weaker. In Section 4.11, we learned that this leads to strong MHD instabilities.

to be spurious, a consequence of MHD instabilities acting like a particle accelerator. The temperatures weren't higher than normal — the instabilities that terminated the discharges were just especially violent. While the claims of the ZETA experimenters were certainly exaggerated by the media, it nevertheless serves as a cautionary tale about communicating fusion results to the public. However, it was more than just that. It was an empirical demonstration of the importance of plasma diagnostics. Arguably as a direct result of this incident, one of the most important diagnostics was developed: *Thomson scattering.*

At the heart of a Thomson scattering diagnostic is a laser, a tightly focused beam of electromagnetic waves with a well-defined wavelength. This laser is fired straight into the plasma. As it passes through, the electrons act like little mirrors, each scattering a small amount of the laser light. Importantly, when scattering occurs, the wavelength of the light is modified in proportion to the velocity of the electron. While this behavior might not be intuitive, the very same thing occurs when a police officer uses a radar gun to bounce an electromagnetic wave off your car. After the plasma scatters the light, a sophisticated camera is used to measure it. By studying the wavelengths it has, we can deduce how fast the electrons are traveling (i.e. the electron temperature). Moreover, the amount of light that is scattered tells us how many electrons there are (i.e. the electron density). Finally, by positioning a number of cameras to view different points on the laser's path (see Figure 5.10), we can measure how the density and temperature change throughout the plasma.

As we will see in the next chapter, soon after Thomson scattering was developed, a tokamak began reporting incredibly high temperatures. This time around, the world was skeptical. They had learned from ZETA. Luckily, Thomson scattering was there to save the day and single-handedly convinced the world that tokamaks were the real deal.

While Thomson scattering remains a workhorse of fusion devices, many more sophisticated diagnostics have been developed. Some can directly measure the small, rapid fluctuations in density and temperature that result from turbulence. Others can map out the

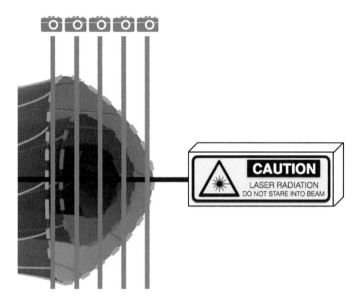

Figure 5.10: The laser path (black line) and the camera lines-of-sight (gray lines) in a Thomson scattering system capable of measuring the density and temperature at five points in the plasma.

magnetic field lines within the plasma. These are invaluable for understanding and improving experimental performance, but many will have to be abandoned in a power plant. In a power plant, we must get by with only a small number of the most robust diagnostics. A power plant will produce so many fusion neutrons that many diagnostics won't be able to survive. Moreover, because the area of the tritium breeding blanket must be maximized, we can't surround the plasma with measurement devices. For these reasons power plants will be relatively blind. Fortunately, they will be guided by decades of pen-and-paper mathematics, enormous computer simulations, and a massive stash of hard-earned experimental knowledge.

5.8 Radioactive Waste and Remote Maintenance

A side effect of fusion technology is the generation of radioactive waste. You may have heard about radioactive waste in the context of *fission* power plants. In a fission plant, the fuel (typically uranium) is broken into two pieces, releasing energy. These two fission products can be radioactive and some take ages to decay into a stable isotope. This material, the highly radioactive byproducts of the reactions that occur inside nuclear reactors, is called *high-level* radioactive waste because it is the most problematic. It continues to generate a significant amount of energy for decades and poses a health risk to humans for tens of thousands of years. Hence, high-level waste must be safely stored for very long periods of time.

The generation of high-level radioactive waste is intrinsic to the generation of fission power. Breaking apart the fuel necessarily creates radioactive material. We stress that, in fusion, this is not the case. A fusion power plant creates no high-level radioactive waste because the product, helium, is stable. The only radioactive waste comes as a side effect of the neutron produced by D–T fusion. Most of these neutrons are absorbed to breed tritium and therefore do not produce radioactive waste.[19] However, the components we position around the plasma (primarily the first wall and the structure of the blanket) will absorb some of the neutrons. For example, when neutrons pass through the first wall sometimes they are captured and transmute atoms into entirely different elements. Many of these elements are radioactive and comprise what is called *intermediate-level* radioactive waste.[20] This, however, is much more manageable. For example, while the US government remains undecided about creating a high-level waste repository, it already has five disposal sites for intermediate-level waste.

Intermediate-level waste decays away more quickly. Specifically, the waste that fusion would produce becomes harmless within a century or so, rather than tens of thousands of years. Additionally, its production isn't intrinsic to fusion power generation. We get to design the components that surround the plasma. We choose which

---

[19] While tritium is radioactive, it is burnt in the plasma and certainly isn't waste (given its monetary value).
[20] *Low-level* radioactive waste typically refers to materials like gloves and tools that have come into contact with intermediate or high-level waste.

materials to use and where to put them. So, as we get better at fusion, we can optimize components to minimize the amount and lifetime of radioactive waste.

Nevertheless, the reactor in a D–T fusion power plant will be far too radioactive for humans to approach, even when the device is shutdown for repairs. For this reason, all maintenance within the vacuum vessel will have to be accomplished remotely using sophisticated robotics. Given the precision and complexity of tokamak components, repairs would be challenging even if humans could do them directly. Not only does this motivate flexible and adept robotic maintenance systems, but also robust and reliable fusion technology. If repairs are difficult, you want to design the machine more conservatively to minimize the number of things that break. Still, even if nothing unexpectedly breaks, portions of the divertor and the blanket will almost certainly need replaced. Robotic maintenance systems have been employed on the JET tokamak and are being further developed by JT-60SA, which is currently under construction.

5.9 Generating Net Electricity

Before we move on, it's important to note that the performance of these technological components can be just as important as the plasma for achieving a viable fusion power plant. At the end of Chapter 4, we introduced the parameter Q, which is the ratio of the fusion power to the external heating power delivered to the plasma. $Q = 1$ corresponds to a plasma that generates as much fusion power as takes to keep it hot and $Q = \infty$ corresponds to an ignited plasma (which requires no external heating). While Q is important for the physics of the plasma (it indicates the fraction of heating that comes from the plasma itself), it isn't the full story. Ultimately, we don't care about the energy balance of the *plasma* — we care about the energy balance of the power plant as a whole. For example, a plasma with $Q = 50$ could still be useless for a power plant, if the heating systems require 100 MW of electricity to deliver 1 MW of plasma heating. On the other hand, if our steam turbine

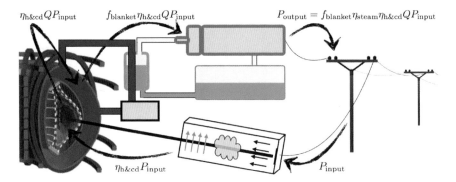

Figure 5.11: The processes that multiply the electrical power used to run a fusion power plant. Here $\eta_{\text{h\&cd}}$ is the efficiency of the heating and current drive systems, f_{blanket} is the power multiplication that occurs in the tritium breeding blanket, and η_{steam} is the steam cycle efficiency of the turbine.[21]

and heating systems were 100% efficient, we could put net power on the grid with $Q < 1$. All these considerations are formalized in the engineering power multiplication factor $Q_{\text{eng}} = P_{\text{output}}/P_{\text{input}}$. This is the ratio of the amount of electrical power a plant puts on the grid P_{output} compared to what it takes from the grid P_{input}. Thus, $Q_{\text{eng}} > 1$ means net electricity production.

For a power plant, Q_{eng} is the fundamentally important parameter and it is expected to be significantly lower than Q. The two biggest reasons for this are inefficiencies in the heating/current drive systems and the steam cycle of the turbine (see Figure 5.11). The heating/current drive systems typically dominate the required external power[22] and the steam cycle determines how much electricity can be generated from the thermal power produced by the reactor. In practice, both of these processes (converting electricity into plasma heating and converting heat in the blanket into electricity) are expected to be about 40% efficient. The only bonus from fusion

---

[21] Everywhere in this figure, Q should be replaced by $(Q+1)$ because the injected heating/current drive power doesn't magically disappear in the plasma. However, we neglect it for simplicity because it will be unimportant as long as Q is large.
[22] The power required to cool the magnets is usually significantly smaller than the heating/current drive power, but is still non-negligible.

technology comes from breeding tritium in the blanket. Instead of generating just 17.6 MeV per D–T fusion reaction, we get an extra 4.8 MeV from the tritium breeding reaction. Multiplying these three conversion factors (i.e. $\eta_{\text{h\&cd}} = 0.4$ from the heating/current drive systems, $\eta_{\text{steam}} = 0.4$ from the steam cycle, and $f_{\text{blanket}} = 1.3$ from the blanket), we see that Q_{eng} is roughly a factor of 5 smaller than the Q of the plasma. Nevertheless, fusion technology presents an opportunity. We see that developing more efficient heating and current drive systems has the potential to substantially improve the viability and economics of fusion. Similarly, developing blankets that can operate at higher temperatures would have the same effect.[23]

For these reasons, quantifying progress in fusion solely by Q or the triple product is an oversimplification. There are a wide variety of ways to improve the prospects of fusion power. In the next chapter, we will discuss some of the breakthroughs that have brought us to where we are today and allude to some of the opportunities we have looking forward.

---

[23] Remember, the steam cycle efficiency is determined by the Carnot limit discussed on page 31. We saw that operating the hot reservoir at a higher temperature enables higher heat-to-electricity efficiencies.

PART 3
THE STATE OF THE ART

Chapter 6

The Past: Fusion Breakthroughs

Throughout decades of fusion energy research, there have been many discoveries and advances — some were intentional and well-crafted, while others were completely accidental. Figure 6.1 shows some of these pivotal moments, each representing a concrete step towards a future fusion power plant. By studying these breakthroughs, one can get a sense for how the field has developed and what sort of progress can be expected in the future. More importantly, the history of fusion is *interesting*. This chapter is, for the most part, a compilation of stories. Stories that are full of unexpected physics, fascinating technology, and bizarre characters — all of which have brought fusion closer to reality.

6.1 1920s: Understanding Stars

Our initial theoretical understanding of fusion stems from the fusion that occurs in nature — in stars. However, even at the beginning of the twentieth century, the mechanism by which stars produced energy was still controversial. Physicists had proposed several theories for how stars produce energy. Two of the most popular were chemical and gravitational contraction.

The chemical theory posited that chemical reactions between elements in the Sun produced the observed solar power. Since both the mass and the power of the Sun were known, it was straightforward to calculate its lifetime, if you assumed the energy density of its

Figure 6.1: Fusion's greatest hits. Future dates are subject to change.

fuel.[1] In 1862, Lord Kelvin, one of the most illustrious physicists at the time, assumed the most energetic chemical reaction known

---

[1]By solving the equation $P = m\epsilon/t$, we can estimate t, the lifetime of the Sun. Here P is the power produced by the Sun, m is the Sun's mass, and ϵ is the specific energy of its fuel (i.e. energy per unit mass).

and calculated that the Sun's fuel could last a mere three thousand years.

The most widely-accepted theory at the time, was that of gravitational contraction. Lord Kelvin thought "that some form of the [gravitational contraction] theory is certainly the true and complete explanation of solar heat." This theory posited that the continual gravitational collapse of the Sun kept it hot. In Chapter 2, we saw that this effect provided the initial heating for the Sun, but calculations revealed that it was only sufficient for a 20 million year life.[2]

These results were completely at odds with other branches of science. Charles Darwin's theory of evolution had put the age of life on Earth in the range of hundreds of millions to billions of years. Since the Sun was observed to power life, it was assumed that the Sun must be older. This timescale found further support from geologists, who established that Earth's oldest rocks were of a similar age. Thus, the physicists were saying that the Sun was at most 20 million years old, while the biologists and geologists said that the Earth was at least hundreds of millions of years old.

We now know that the Sun is roughly 5 billion years old. So why did physicists get it so wrong? The main culprit was their lack of knowledge of the nucleus and, therefore, of nuclear power. This story provides a remarkable demonstration of the difference in the energy density of chemical bonds versus nuclear bonds. Kelvin showed that a Sun's worth of chemical reactions gets you a few thousand years at most. However, a Sun's worth of fusion reactions gets you a few billion years.

The discovery that paved the way for understanding fusion and its place in the cosmos was Einstein's statement of the equivalence of mass and energy: $E = mc^2$. The next ingredient was a stupendously brilliant discovery by Francis William Aston, an English physicist

---

[2] To calculate the solar lifetime, Hermann von Helmholtz, a distinguished German physicist, simply calculated the Sun's total gravitational potential energy and divided it by the Sun's total power.

and chemist. As is often the case in science, this discovery was serendipitous. While Aston was looking for isotopes of neon, he noticed that four hydrogen-1 nuclei were somewhat heavier than a single helium-4 nucleus. In 1920, the prominent English astrophysicist Sir Arthur Eddington put two and two together[3] and hypothesized that this could be the origin of the Sun's energy production. If all of the hydrogen in the Sun were converted into helium, the energy of the extra mass, according to $E = mc^2$, could support the Sun for around 100 billion years. Suddenly, the estimate of the lifetime of the Sun had increased from thousands of years to one hundred billion years.[4]

Now that the basic mechanism for stellar energy production had been identified, physicists set to work calculating exactly which isotopes could be fused and how much energy they released. In 1938, the master of nuclear physics, Hans Bethe, synthesized all of this information into a trailblazing paper, *Energy Production in Stars*.[5] In one fell swoop, he systematically classified most of the possible fusion reactions in stars and how they depend on the star's mass (which determines its temperature).

Bethe found that, in small stars (i.e. less than 1.3 times the mass of our Sun), a set of fusion reactions called the proton–proton chain dominates. This reaction is shown on the left side of Figure 6.2 and essentially takes four protons and directly converts them into one helium-4 nucleus (just as Eddington surmised). Each such set of reactions releases 26.7 MeV in total, even more energy than D–T fusion. In more massive stars, Bethe found that the carbon–nitrogen–oxygen (CNO) cycle dominates (shown in the right side of Figure 6.2). This remarkable set of reactions again takes four protons and turns them into a helium-4 nuclei (also releasing 26.7 MeV of

---

[3] No pun intended.
[4] The Sun's lifespan is substantially shorter than Eddington's estimate of 100 billion years because the Sun won't use up all of its fuel before it dies.
[5] If you have an undergraduate physics background (or equivalent), the paper is highly readable [24].

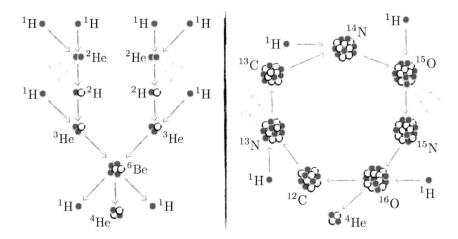

Figure 6.2: The proton–proton chain (left), which dominates energy production in smaller stars like our Sun, and the CNO cycle (right), which dominates in larger stars.

energy). However, the CNO cycle differs markedly from the proton–proton chain in that it requires the heavy nuclei of carbon, nitrogen, and oxygen to serve as catalysts.

In 70 years, humans went from thinking that chemical reactions might power the stars to a fully-developed theory of the nuclear processes underlying stellar physics.

6.2 1950s: A Kick-Start for Fusion

Following World War II, scientists took fusion research from the stars and began work on bringing it down to Earth. Initially, the only application was nuclear weapons — specifically to follow-up on the Manhattan project and build "super," the hydrogen bomb. However, on March 24, 1951, a little-known scientist named Ronald Richter in Argentina changed this all.[6] On this date, he and the Argentinian president Juan Perón held a press conference claiming that they had achieved "enormous temperatures of millions of degrees" and

---

[6]For a more detailed and personal account of Ronald Richter's story we recommend *Atomic Adventures* by James Mahaffey.

produced the "controlled liberation of atomic energy" in a "solar reactor apparatus" called the "thermotron." The thermotron, they said, would be employed "solely for power plants, smelters and other industrial establishments."[7] This made headlines around the world. The front page of the New York Times proclaimed "PERON ANNOUNCES NEW WAY TO MAKE ATOM YIELD POWER." However, while Richter and Perón made grand claims, they didn't back them up with much technical detail. Hence, many of the world's leading nuclear scientists were asked to comment on the results with no specific methodology to critique. In other words, they were basically asked "Is fusion power generation feasible?"

In the months that followed, Richter was revealed be as close to a mad scientist as exists in the real world. After fleeing from post-war Germany, he had convinced Perón of his vision to accomplish fusion using electric arc discharges. He then burned through nearly $300 million dollars (adjusted to 2017 dollars) in constructing an elaborate fusion laboratory on Huemul Island, a tiny island in a lake deep within Patagonia. Shrouded in secrecy and protected by the army, he experimented with proton-lithium fusion and quickly became convinced of his own success. Unfortunately, his fall came as quickly as his rise. The worldwide attention prompted Perón to properly scrutinize the thermotron project for the first time. More level-headed Argentinian scientists were assembled to perform an inspection, which resulted in a scathing condemnation of Richter. A year later, the project was quietly abandoned along with Richter himself.[8]

But the story doesn't end there. Richter's dramatic pronouncement stimulated the scientific community around the world. Lyman Spitzer, an astrophysicist at Princeton University, was preparing for a ski trip when he received a phone call from his father, who had just read the New York Times article. While Spitzer dismissed the thermotron, he spent his trip brainstorming alternatives. When he

---

[7]They also claimed that all scientists working on the hydrogen bomb were "enormously far from their goal." I find the complete and utter audacity of this whole situation impressive.
[8]No, he wasn't killed, just briefly imprisoned.

returned home, he submitted a grant proposal to the US Atomic Energy Commission and was awarded $50,000 to build his most promising idea: the stellarator. Emboldened by this, James Tuck, a British scientist at Los Alamos, immediately began work on his own concept: the Perhapsatron. Richter's story even penetrated the Iron Curtain, animating the project that would become the tokamak, which had been bogged down in Soviet bureaucracy. Undoubtedly, these three projects encouraged the funding of fusion around the world, if for no other reason than to prevent a fusion energy "gap."

6.3 1960s: Superconducting Magnets

As evidenced by the name, magnetic confinement fusion devices require very strong magnetic fields. Generating these fields with traditional conductors, like the copper used to wire your house, would consume enormous amounts of power. So much in fact, that magnetic confinement fusion would look pretty impossible. For this reason, superconductivity is a godsend. As we learned last chapter, if we drive electric current through a regular conductor, it will encounter resistivity and continually drain away the current into heat. However, in a superconductor, the current (and the magnetic field it creates) will persist indefinitely because there is no resistance.[9] Superconductivity was first discovered in 1911, when Dutch physicist Heike Onnes cooled mercury to 4 Kelvin (which is 4°C above absolute zero or $-269°C$).[10] Soon after, superconductivity was observed in a number of other materials, yet it evaded theoretical explanation for over 50 years.

So why is superconductivity possible? The short answer is that it is a subtle consequence of quantum mechanics. While quantum

---

[9]Superconductors have sustained current for years with no measurable sign of degradation.

[10]Mercury was particularly easy to use experimentally since it was straightforward to obtain pure samples, an important consideration in the early twentieth century. However, the enabling factor for these experiments was the liquidification of helium by Onnes in 1908, which allowed these incredibly low temperatures to be reached.

mechanics leads to all sorts of unintuitive behavior, normally it is relegated to the subatomic world. Superconductivity is one of the few ways that it manifests on human scales. The long answer is more technical, so it is confined in the following Tech Box. Fortunately for us, the reason that superconductors were such a breakthrough for fusion is more to do with their technological development, rather than the physics underlying them. Obviously, resistivity-free magnets are immediately attractive, but, on its own, this isn't enough to be useful for fusion.

A superconducting material has three limits: current-carrying capacity, temperature, and magnetic field. If you try to put too much electric current through a superconducting coil, operate it at too high of a temperature, or in too strong of a magnetic field, it will lose its superconductivity. For early superconductors like mercury, these limits were extremely restrictive (see Figure 6.3). The main problem was that the maximum magnetic field they could tolerate was around

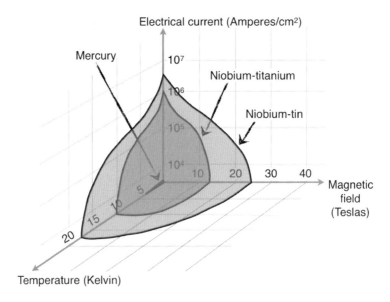

Figure 6.3: The limits of three different superconducting materials: Niobium–tin, niobium–titanium, and mercury (barely visible). If the material is operated in conditions that are *below* the surface shown in the plot then it is superconducting, otherwise it behaves as a normal conductor. It is immediately apparent that the two niobium compounds are much more useful than mercury.

0.05 Tesla! Remember, for fusion, we need fields exceeding 1 Tesla. Hence, for all practical purposes the early superconductors were useless (as concluded by Onnes in his own Nobel prize acceptance speech).

TECH BOX: Why is superconductivity possible?

Normally, the electrons that carry electric current repel one another because they are each negatively charged. However, when moving in the rigid lattice of a solid material, an electron attracts the surrounding positively charged nuclei. This is similar to quasineutrality in a plasma (see Figure 4.5 on page 98), except the accumulation of positive charge from the lattice deformation can actually exceed the negative charge of the electron. Hence, the net build-up of positive charge can attract a second electron and it can become bound to the original one. The binding between the two electrons is very weak, so it is destroyed by thermal vibrations in the lattice unless the temperature is extremely low.

These pairs of electrons are called *Cooper pairs* and the number that can stably form in a material determines the superconductor's current-carrying capacity. Each pair of electrons necessarily has opposing quantum mechanical spin and acts very differently than individual electrons. Specifically, it is no longer required to follow the Pauli exclusion principle. Hence, all the Cooper pairs can become quantum mechanically correlated and act as one single fluid. This is the key point because it makes the flow of electric current robust against the electron–nucleus collisions that cause resistivity. Normally, when an individual electron hits a nucleus, the collision contains sufficient energy to kick the electron into a different quantum mechanical state (e.g. one that is no longer moving along with the rest of the electric current). However, in the fluid of Cooper pairs, all of the electrons are tied together, so

(*Continued*)

(*Continued*)

> any collision with a nucleus would have to provide sufficient energy to kick *all* of the electrons into a different energy state. Since collisions are not that energetic, the flow of Cooper pairs is entirely unaffected by collisions and the electric current avoids resistivity.

The first inkling of practical superconductivity came in 1935, when Ukrainian scientists Rjabinin and Schubnikow discovered a material that exhibited a new kind of superconductivity. These materials were inventively named *type II* superconductors in order to distinguish them from the existing superconductors, which became known as *type I*. While type I materials suddenly and entirely lose their superconducting ability when any of the three operating limits is violated, type II materials exhibit a gradual transition. At the start of this transition, only small regions in the material lose superconductivity, which the electric current can simply go around. This means that type II superconductors remain resistivity-free over a much broader range of operating conditions. Early type II superconductors were used to construct coils that generated 1 Tesla magnetic fields. While this was encouraging, the real breakthrough for fusion came in 1961. Scientists at the Bell Telephone Laboratory discovered that a niobium–tin compound, a type II superconductor, was capable of functioning in fields exceeding 10 Tesla (see Figure 6.3). Niobium–tin, along with a similar niobium–titanium compound, have since become the workhorses of applied superconductivity. Niobium–titanium is cheaper and more robust, but niobium–tin can achieve higher magnetic fields.

Progress in superconductivity is enormously useful for fusion, but there is typically a substantial delay between discovery and implementation (see Figure 6.4). Every step from the discovery of a superconducting material, to producing macroscopic quantities of it, to engineering mechanically strong wires, is a challenge that holds

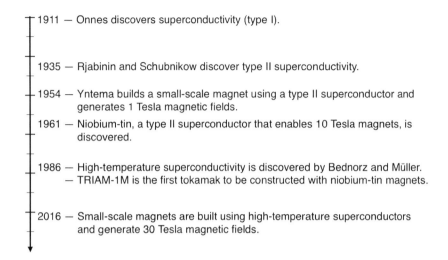

Figure 6.4: A brief history of superconductivity, which shows the timescales required to develop a scientific breakthrough into a practical technological advance.

the possibility of failure. Hundreds of superconducting compounds have been identified, yet only a few are practical. So it isn't surprising that the first fusion device to use superconducting magnets, the T-7 tokamak at the Kurchatov Institute in Moscow, wasn't built until 1979 — almost 70 years after the discovery of superconductivity. It used toroidal field magnets made of niobium–titanium to generate a 3 Tesla field. The first tokamak to use the more ambitious niobium–tin superconductor was TRIAM-1M in Japan, which had an impressive field strength of 8 Tesla. However, it wasn't built until 1986, 25 years after the superconducting properties of niobium–tin were first observed.

While the length of this technological delay is unfortunate, it does allow us to anticipate the future. More exciting advances for fusion magnets may be just around the corner! In 1986, Bednorz and Müller discovered the first *high-temperature* superconductor, which is defined by permitting operation above 30 Kelvin. However, the real benefit of high-temperature superconductors isn't their operating temperature, but rather that some can sustain magnetic fields exceeding 100 Tesla! Moreover, unlike the old materials, high-temperature superconductors can be built with joints, enabling them to be disassembled. This would revolutionize the maintenance schemes possible in a tokamak. Currently, components like the vacuum vessel must be cut into toroidal sections, so they can fit through the gaps between the toroidal field coils. Though this might seem mundane, the ability to install whole components "plug & play"-style would make a fusion power plant much more practical. Research into high-temperature superconductors is ongoing. Just in the last couple of years, Rare Earth Barium Copper Oxide (REBCO) coils have been used to generate 30 Tesla magnetic fields, albeit in small-scale magnets. As we will see in Chapter 9, two of the fusion startups are already in the process of employing REBCO high temperature superconductors. So we see that, even if the fusion community is unable to make any progress, it can still be swept towards a practical power plant by progress in other fields like superconductivity.

6.4   1960s: The Tokamak

In 1950, a 23 year-old Soviet soldier named Oleg Lavrentyev was stationed on Sakhalin, a remote Russian island north of Japan. With nothing more than an eighth-grade education and access to an army library, he made himself into a self-taught expert on fusion. In July of that year, he sent a letter to the Kremlin proposing two concepts: a rough schematic for a hydrogen bomb and a fusion power plant design based on *electrostatic* plasma confinement. His letter reached Andrei Sakharov, a leading scientist in the Soviet nuclear weapons program.[11] After a few months of careful consideration, Sakharov and a colleague, Igor Tamm, improved upon Lavrentyev's idea. They

---

[11] Contrary to what you might think when you hear "leading scientist in the Soviet nuclear weapons program," Sakharov was an inspirational person. In 1975, he was awarded the Nobel Peace Prize for his efforts in advocating nuclear disarmament and defending civil liberties in the Soviet Union.

submitted a proposal to the Soviet government to begin a *magnetic confinement fusion program.*[12]

For the better part of a year, Sakharov's proposal gradually made its way through the bureaucracy of the Communist Party. Coincidentally (or not?), shortly after Ronald Richter's supposed accomplishments hit the newspapers, the project got Stalin's signature and as well as great priority. The Soviets budgeted money for experiments, transferred eminent scientists from the hydrogen bomb program, and allocated the best students as they graduated from university. Clearly, the Soviets considered fusion research to be of substantial importance, which is likely the reason they led the field for so long.

At this point, fusion efforts around the world were completely classified. Cold War tensions were high and governments still believed that controlled fusion could have near-term military applications. The first step towards declassification was made in a surprise move by the Soviets. In 1956, their leader Nikita Khrushchev led a delegation to England that included Igor Kurchatov, the director of the Soviet atomic bomb program. As part of the trip, Kurchatov gave a lecture at the United Kingdom's Atomic Energy Research Establishment at Harwell. His talk floored the British, both for its frank and open discussion of what was considered classified material and for how advanced Soviet fusion research was. Ironically, a significant part of the talk was devoted to the difficulty of distinguishing between neutrons generated by high temperatures (which is the goal) and neutrons generated by plasma instabilities (fairly useless). Just a year later, a lack of appreciation of this subtlety led the British to inaccurately proclaim the success of ZETA. Kurchatov's lecture

---

[12]Sakharov's proposal had the goal of a 5 Tesla D–D power plant that produced almost a gigawatt of fusion power. The size he proposed, a donut with a 12 meter radius and a 2 meter cross-sectional radius, is similar to modern-day estimates for a D–D power plant. However, this preliminary design included a magnetic coil in the center of the plasma. Sakharov did consider using the plasma to carry the current, but recognized that electromagnetic induction could not sustain it in steady state.

galvanized the complete declassification of controlled fusion research around the world, which would be vital in enabling the rapid progress of the tokamak in the coming decades.

Soviet scientists built the first true tokamak, T-1, in 1958. However, the early experiments were plagued by heavy impurity atoms (e.g. iron and carbon), which diffused out of the first wall during experiments. These impurities ended up in the plasma and, because of their high electric charge, radiated away large amounts of

Oleg Lavrentyev was a soldier and self-taught fusioneer who inspired the Soviet controlled fusion program.

Andrei Sakharov who, along with Igor Tamm, invented the tokamak.

Igor Kurchatov, also known as "The Beard," was the director of the Soviet atomic bomb program. He made a pledge not to cut his beard until the program succeeded.

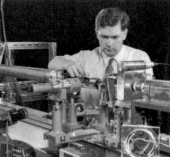

Michael Forrest, a British experimental physicist, with the Thomson scattering diagnostic he helped to develop and install on the T-3 tokamak.

Figure 6.5: Some of the early pioneers of the tokamak.

the thermal energy and ruined the confinement time.[13] Eventually, through fastidious cleaning and the development of the limiter (see page 158 for a reminder), the impurities were minimized.[14]

The result was that, in 1968, diagnostics on the Russian T-3 tokamak began reporting incredible performance — temperatures in excess of 1 keV! Having learned from ZETA, the Russians invited a team of British scientists to confirm their results using one of the world's first Thomson scattering systems. The British scientists, in the height of the Cold War, packed up their instruments and hopped on a plane to Moscow.[15] This act of international scientific collaboration was not only heartwarming, but also confirmed that the results were real. Overnight, the fusion community was changed forever. Skeptics became supporters and, within six months, the Model C stellarator at Princeton was converted into the Symmetric Tokamak. The age of the tokamak had begun.[16]

6.5 1970s: Bootstrap Current

To enable a tokamak power plant to operate in steady state, we must rely on non-inductive current drive techniques like neutral beams and

---

[13] The plasma isn't hot enough to remove all of the electrons from heavy impurity atoms. The thermal energy of the plasma repeatedly excites these bound electrons to higher atomic energy levels and, when they decay back to their ground state, they release a lot of electromagnetic radiation.

[14] Tokamaks are cleaned by repeatedly heating the first wall up to a high temperature, which tends to eject any loose particles from the surface. Additionally, plasma discharges are performed without any confining magnetic fields, causing the plasma to knock loose particles out of the wall. Pretty sophisticated stuff, right? It's not like they send graduate students into the device with Windex® and paper towels.

[15] For a detailed and first-hand account of the trip, we recommend *Lasers Across the Cherry Orchards* by Michael Forrest.

[16] Oleg Lavrentyev passed away in 2011. After inspiring the Soviet controlled fusion program, he enrolled in the physics program at Moscow State University and was sponsored by Lavrenti Beria, the chief of the Soviet secret police. When Stalin died in 1953, Beria fell out of favor and was executed. Unfortunately, because people associated Lavrentyev with Beria, he was ostracized. However, he forged his own path. Lavrentyev moved to the Ukraine and led a productive research career, ever focused on electrostatic plasma confinement.

electromagnetic waves. While both of these methods can sustain the plasma current indefinitely, they typically require lots of power to do so. This is bad, since it can consume a significant fraction of the electricity that we would otherwise sell to the consumer. Almost unbelievably, in this instance, the plasma itself comes to the rescue. In the early 1970s, theoreticians in Russia and the United Kingdom, through pencil and paper math, discovered that the plasma can intrinsically generate its own current. A decade or two later, this current was confirmed by experimental observation. This mechanism was named the *bootstrap current* because it enables the plasma to "pull itself up by its own bootstraps." Ideally, all we must provide is a small "seed" current to get things going, then the plasma will multiply it many times over via the bootstrap effect. So, how does this fortuitous phenomenon work?

To understand, we need to revisit some plasma physics, specifically trapped particles and banana orbits. In Chapter 4, we learned that trapped particles get reflected due to the magnetic mirror force,

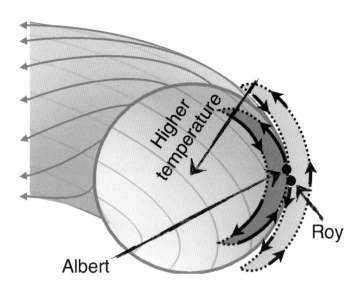

Figure 6.6: The trajectories of two trapped electrons, named Albert and Roy, projected onto a poloidal plane. As they execute banana orbits, they experience different temperatures, giving rise to a bootstrap current that points out of the page.

making them follow banana orbit trajectories (shown in Figure 6.6). We found that the trapped particles reduced the confinement time because banana orbits allowed them to travel further away from their magnetic surface. Well, now we will see that the large width of the banana orbits has another effect, but this time it's helpful.

In fusion devices, the hottest point in the plasma is at the center. This makes sense because it is furthest away from the solid material at the edge (which must be cool so as not to melt). This temperature gradient means that, when trapped particles drift across magnetic surfaces as part of their banana orbits, they encounter different temperatures. If they drift inwards, the temperature increases; if they drift outwards, the temperature decreases. While we've relegated the details to the Tech Box below, this actually leads to a spontaneous electric current. Electrons moving in one direction along the field line drift inwards and experience higher temperatures (on average). This means that they move faster (on average) than the particles moving in the opposite direction on the same field line. Hence, the electrons have net motion and an electric current arises.

TECH BOX: The bootstrap current

To understand the bootstrap current, we will follow two trapped electrons that start from the same point, but move in opposite directions along the magnetic field line (as shown in Figure 6.6). Let's call them Albert and Roy after two of the scientists who predicted the bootstrap current. Initially, Albert is traveling along the field line in the direction going into the page, while Roy is traveling along the same field line, but moving out of the page. Both Albert and Roy drift downwards due to the ∇B and curvature drifts, which means that Albert drifts inwards (towards the center of the plasma) and Roy drifts outwards (towards the edge of the plasma). Since Albert spends more of his time closer to the center of the plasma, he will experience hotter temperatures and be

(Continued)

(Continued)

> moving faster than Roy on average. This gives rise to a current because, when we add up the motions of all of the electrons on the magnetic surface, Albert and Roy don't cancel. Albert is hotter than Roy, so he moves faster and carries more current. This leads to a net current that comes out of the page, which reinforces the preexisting plasma current.[17] The preexisting "seed" current is necessary because otherwise there will be no poloidal field and, hence, no trapped particles or banana orbits.

In this way, much of the current needed in a tokamak can be provided by the bootstrap mechanism. In some power plant designs, the bootstrap current is a factor of ten larger than the externally driven current. This, however, is ambitious. In order to maximize the bootstrap effect, we must optimize the variation of temperature (and particle density) throughout the device. Unfortunately, doing so also has an effect on the plasma confinement. Since the best configuration for the bootstrap current isn't necessarily the best for the confinement time, the optimal solution isn't always clear. Nevertheless, even a modest bootstrap current significantly improves the prospects of a fusion power plant. Driving the plasma current with neutral beams or electromagnetic waves requires a lot of power. Even if the bootstrap mechanism were to cut the current we needed to drive in half, it would make it much easier to produce net electricity. For the first time, the plasma decided to cut us some slack and reward us for our hard work!

---

[17]The motions of the ions can be ignored because they are so heavy that they move slowly compared to the electrons. Since electrons have a negative charge, they create an electric current in the direction that is opposite to their motion. Also, the bootstrap current is further enhanced because the trapped particles collide with non-trapped particles and drag them along somewhat. The non-trapped particles that are dragged by the trapped particles actually carry most of the bootstrap current.

6.6 1980s: H-Mode

On February 4, 1982, scientists working on the ASDEX tokamak in Germany noticed something remarkable. During a routine neutral beam heating experiment, the plasma underwent a sudden transition in performance. The energy confinement time τ_E doubled and the plasma formed a remarkable, turbulence-free zone near the edge. This zone became known as the *pedestal*. As a result of the transition, the gradients of temperature and density in the pedestal increased significantly, thereby raising the temperature and fusion power of the entire core (see Figure 6.7). This new operating mode was named *H-mode* for high-mode and the old way of operation became known as *L-mode* for low-mode. The lead scientist behind the experiments, Fritz Wagner, quickly mobilized an effort to spread the Good News. "H-mode" t-shirts were printed and a lecture tour was planned. As usual, the fusion community greeted the dramatic pronouncement with an ice-cold bucket of skepticism. However, in the following years, H-mode operation was achieved on nearly every other tokamak around the world. Eventually, it was even observed in a stellarator, showing it to be a robust characteristic of toroidal plasma confinement devices. Now, 35 years later, it is

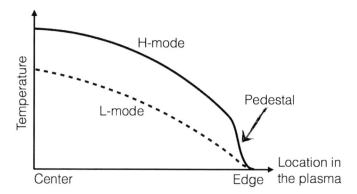

Figure 6.7: A comparison of the temperature throughout the poloidal cross-section of an H-mode and L-mode plasma. Notice the primary difference is the presence of a narrow "pedestal" at the edge, which raises the temperature of the entire H-mode plasma.

hard to imagine designing a power plant that doesn't operate in H-mode.[18]

The key to achieving H-mode is to exceed a certain threshold of heating power. A plasma discharge starts out in L-mode, but if you blast it with enough neutral beam and electromagnetic wave heating, the turbulence near the edge will be killed off and the plasma will jump into H-mode. The reason it took so long to discover this phenomenon is that the critical amount of heating can be prohibitively large. Fortunately, there are some simple experimental steps that can be used to reduce it to something practical. For example, the power threshold for H-mode is lowered when

- the plasma has a divertor (rather than a limiter),
- deuterium is used (rather than normal hydrogen),
- the first wall is very clean (reducing impurities in the plasma), and
- the plasma cross-section has certain shapes (e.g. just one X-point with the ion ∇B particle drift directed towards it).

While the H-mode discovery brought unprecedented confinement to tokamaks, it didn't come without a price: *edge localized modes* (ELMs). ELMs are periodic explosive bursts of particles and energy from the plasma edge. They are a unique feature of H-mode and play a large role in determining the plasma confinement. In L-mode, turbulence at the plasma edge happily bubbles away, continuously transporting particles and energy. In H-mode, turbulence is strongly stabilized, which means that the density and temperature in the edge gradually build up. Eventually some local MHD stability limit is crossed and the plasma belches,[19] expelling a large burst of plasma (see the rightmost image in Figure 6.8). Remember, way back in Chapter 4 we learned that MHD (i.e. magnetohydrodynamics) models the plasma as an electrically charged fluid. While crude, this model can still predict the onset of some large-scale plasma instabilities — ELMs are one such example.

---

[18]The original H-mode paper is Ref. [25].
[19]You can also think of it a burp ... or a fart. Really the human body is full of disgusting ELM analogies.

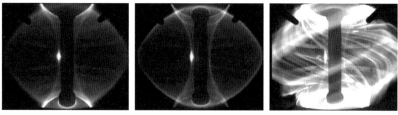

Courtesy of the UK Atomic Energy Authority

Figure 6.8: A front view of the MAST tokamak with an L-mode plasma (left), an H-mode plasma (center), and an H-mode plasma during a burst of ELMs (right). Notice how the L-mode plasma has a fuzzier boundary than the H-mode plasma.

While the overall H-mode confinement (including ELMs) is still substantially better than L-mode, this behavior is concerning. The fact that the energy comes out in large bursts poses a danger to the first wall and divertor. In a future power plant, ELMs are expected to contain enough energy to significantly shorten the lifetime of plasma-facing components, which would increase maintenance costs and reactor downtime. Additionally, when ELMs hit the first wall, they can cause the release of impurities, which negatively affects plasma performance. Fortunately, some strategies have been developed to reduce the energy of ELMs and/or eliminate them entirely.

TECH BOX: Combating ELMs

ELMs have been partially suppressed in many experimental devices, generally by one of two techniques. The first is to shoot small frozen pellets of solid deuterium into the plasma edge, a technique known as ELM pacing. Each pellet gives the plasma a bit of a kick, which can trigger the ELM prematurely and increase the frequency of ELMs. While this might sound like we've made the problem worse, it's actually better — since the ELMs are occurring more frequently, the amount of energy each releases is lower. This way, we have lots of smaller ELMs,

---

(*Continued*)

(*Continued*)

rather than a few big ones. The second ELM suppression method is to perturb the plasma edge with weak magnetic fields that are not toroidally symmetric. This creates a thin stochastic region, which worsens confinement. Effectively, we are manually degrading the confinement such that the ELMs don't step in and do it themselves. Charged particles can then flow out of the device along the stochastic field lines, releasing the energy gradually, instead of in large ELMs.

Of course, both ELM suppression techniques are undesirable because they would increase the complexity of a power plant and reduce confinement somewhat (though it still remains significantly better than L-mode). For this reason, it is very exciting that scientists have observed instances of H-mode operation without ELMs: the so-called *quiescent* H-mode as well as *I-mode*. Although it's unclear how universal these modes of operation are, they hold the potential for fusion power plants with H-mode confinement times, but without damaging ELMs.

H-mode is also fascinating from a theoretical perspective. Suppressing turbulence is the holy grail of fusion and we've just learned that it takes place in the narrow pedestal at the edge of an H-mode plasma. So is there a way to make this lovely pedestal extend all the way to the center of the plasma? If this were the case, the confinement time in our devices could approach the neoclassical estimate (i.e. very good) and a fusion power plant would be dramatically more achievable. The short answer is that we're not sure.[20]

The community does not have a first-principles mathematical framework that is capable of predicting the width of the pedestal

---

[20]Experiments have created so-called *internal transport barriers*, which resemble the H-mode pedestal and are located in the plasma interior. However, they are a much less robust phenomenon.

and the amount of heating power needed to trigger its creation.[21] Considering the practical importance of H-mode, this is a remarkable fact. It is a testament to the sheer theoretical complexity of the problem. Some of the smartest people in the world have been banging their heads against it for decades using the most powerful computational tools available. Yet the problem remains.

That being said, some theoretical progress has been made. Gyrokinetic simulations (which we will discuss later in this chapter) have revealed the importance of *plasma flow* in stabilizing turbulence. Plasma flow refers to the bulk velocity of the plasma. When the flow velocity varies significantly from magnetic surface to magnetic surface, it can "shear" turbulent eddies before they can grow to be big and strong (see Figure 6.9). This works because different parts of an eddy are moving at different speeds, so the whole thing gets torn apart like a dollop of milk being stirred into coffee. In Chapter 2, we saw that air flow driven by the Earth's rotation, does the same thing to the large convective eddies in the atmosphere. Although important details remain unresolved, the general consensus is that plasma flow

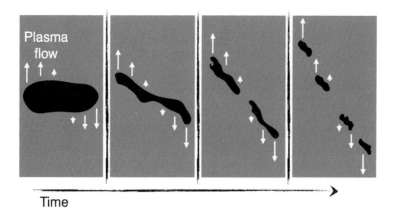

Figure 6.9: A four-panel cartoon of a single turbulent eddy that is broken up into smaller, weaker eddies by a spatial variation in plasma flow.

---

[21] We should mention that some models show great promise. For more information on the leading predictive pedestal model, EPED, see Refs. [26, 27].

shear plays an important role in quenching turbulence in the H-mode pedestal.

6.7   1980s: Plasma Shaping

While the magnetic geometry of a tokamak is much more constrained than a stellarator, there is still considerable freedom. The plasma must remain symmetric in the toroidal direction, but poloidal field coils can be used to create a wide range of cross-sectional shapes (see Figure 5.2 on page 143 for a reminder). Initially, all tokamaks had circular cross-sections because ... well ... why not? It seemed like a good place to start. The big breakthrough came in 1973, during the design of the JET tokamak. While JET is noteworthy for being the world's largest tokamak, it was also the first device designed to create a "D"-shaped plasma. From a modern perspective, this was an incredibly gutsy call.[22] Big fusion experiments are so expensive that they tend to be very conservative. JET cost hundreds of millions of dollars to build and was funded by a consortium of European governments. In other words, it was the last machine you would expect to voluntarily foray far into the unknown. Moreover, at that point, there was scant evidence motivating shaping at all, let alone a "D"-shape in particular.[23] So why did the designers of JET make such a bold decision? For reasons largely unrelated to plasma physics.

A few years earlier, fusion researchers at Princeton had worked on optimizing the shape of the *toroidal field coils*. They found that some shapes are better than others at handling the electromagnetic forces that the magnet exerts on itself. Using these shapes allows coils to generate the same magnetic field with less structural reinforcement, making them less expensive. So which shape did researchers find to

---

[22] If the "D"-shape turned out really bad, JET did still have the ability to create a circular plasma, but it wouldn't have been a particularly efficient or optimized design.

[23] At this point, there had been a series of successful "doublet" experiments at General Atomics, which explored a highly elongated plasma shape (shown in Figure 6.11) that included a magnetic X-point in the center of the plasma.

be the best? You guessed it, a "D."[24] The designers of JET used this shape for their toroidal field coils and reasoned that, since they are the most expensive part of the device, they might as well use them as efficiently as possible. Hence, in order to best utilize the available space, they gave their first wall and plasma the same shape. In hindsight, this makes a lot of sense. Doing anything else would be like trying to put a square peg through a round hole — it's possible, but requires the peg to be much smaller than the hole. Thus, JET made its plasma "D"-shaped in order to maximize the total amount of plasma that would fit within "D"-shaped coils.

What was unclear at the time was how the new shape would affect the plasma behavior. Researchers calculated that changing the shape from circular should allow more current to be driven in the plasma without triggering MHD instabilities.[25] However, the effects of more current were largely unknown. Moreover, everyone was completely clueless about the effect of the new shape on turbulence. So how did JET's gamble pay off?

Well, everyone had to wait quite a while to find out. After the design of JET was mostly completed, the project stalled. As one of the earliest international scientific experiments, it got bogged down in politics. Specifically, the different European governments couldn't agree on where to put it. The field had been narrowed down to just two locations, either near Oxford in the UK or near Munich in Germany, but the situation became a stalemate ... for over a year and a half.

Then, at 11 am on October 13, 1977, Lufthansa Flight 181 was hijacked by terrorists after departing from Spain. This was a

---

[24]This is a consequence of the fact that the magnetic field is strongest on the inboard side of the device (see page 108 for a reminder), so the electromagnetic forces are stronger there. Having the vertical flat part of the "D" on the inside allows the central solenoid to provide additional support for the toroidal field coils.

[25]This is because a circle is the shape with the smallest perimeter to area ratio. Making the plasma cross-sectional shape non-circular increases the length of the perimeter, thereby increasing the safety factor q at constant plasma current. This means you can either choose to let the plasma become more stable or to drive more current (maintaining the same level of plasma stability).

major escalation of the German Autumn, a series of attacks on West German targets by pro-East German groups. The terrorists, seeking the release of twelve of their compatriots and $15 million, took the plane on a 10,000 kilometer journey that lasted nearly five days. The world could only watch in horror as the terrorists directed the plane from Italy to Cyprus to Bahrain to the United Arab Emirates to Yemen, where they killed the plane's captain. Finally, in Somalia, German special commandos stormed the plane using specialized stun grenades provided by the UK. The operation was a complete success, killing or disabling all the terrorists with very minimal casualties to the commandos and hostages. In the emotional aftermath, the German chancellor, grateful for the assistance provided by the UK, took steps that allowed JET to be built in England.[26]

With the location finally selected, construction could finally begin. Five years later (both on time and on budget!), JET began to produce results. They were overwhelmingly positive. First, as anticipated, shaping the plasma did indeed allow JET to drive more plasma current, by about a factor of two. Importantly, this extra current was found to roughly double the confinement time. Moreover, shaping turned out to have an additional, direct effect on turbulence. The "D"-shape was found to increase the confinement time by an additional 50%. This direct reduction in turbulence was a complete surprise to the community and is still not fully understood. Nevertheless, there is no doubt that the "D"-shape was instrumental in JET's subsequent record-breaking performance (as we will see in the next section).

A great illustration of the success of JET's shaping is shown in Figure 6.10. We see that, after JET began operations, nearly every new tokamak adopted a "D"-shaped plasma cross-section. However, this is somewhat ironic, since, even after JET's gamble paid off, later tokamaks were unwilling to explore shaping much further.[27]

---

[26] A detailed account of the politics of JET (including more details on this story) can be found in *A European Experiment* by Denis Willson.
[27] Of course, these tokamaks vary in many other important ways (e.g. heating systems, divertors, diagnostics).

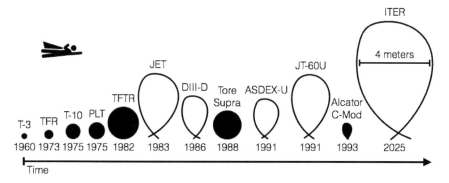

Figure 6.10: The evolution of the cross-sectional plasma shape from T-3, one of the first tokamaks, to ITER, which is due for completion around 2025. Note that, after the construction of JET, tokamaks became overwhelmingly "D"-shaped. The tokamaks represented by the filled black silhouettes are no longer operating. RIP.

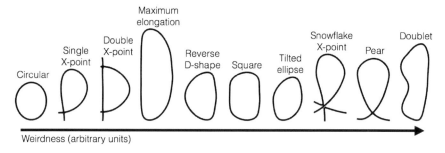

Figure 6.11: A selection of the plasma cross-sectional shapes that have been achieved in TCV, many of which either hold great promise or illuminate the physics that occur in more conventional shapes. They are organized from most conventional to weirdest (as judged by us).

For this reason, the Swiss built a tokamak in 1992 specifically to investigate novel plasma shapes: Tokamak à Configuration Variable (TCV). TCV includes a very tall vacuum vessel as well as a bunch of high-capacity poloidal field coils, which allows it to explore a wide variety of shapes that are inaccessible to other devices (see Figure 6.11).

6.8 1990s: Deuterium–Tritium Fuel

In this book, we have almost exclusively focused on the deuterium-tritium fusion reaction. This is for good reason — in Chapter 3 we learned that it is substantially easier than any other fusion reaction. Thus, it seems like large tokamak experiments would always be fueled with a 50–50 mixture of deuterium and tritium. This would enable them to maximize the fusion power and most accurately mimic the conditions of a future power plant. In reality, D–T tokamak plasmas have only been created in three sets of experiments ever — and all of them took place in the 1990s. While tritium would enable devices to produce significant amounts of fusion power, using it is an expensive headache.

Instead of D–T, fusion experiments generally use pure deuterium fuel. This is done for several reasons. First, tritium costs about $10,000 per gram, while a gram of deuterium only costs about $1. Second, the costs don't end after you buy it. Tritium is radioactive, which means it must be transported and handled according to strict regulations to ensure that it is never released to the environment. Moreover, having radioactive material on site necessitates stricter security and safety standards.[28] However, the most important reason is that the D–D fusion reaction is much *less* likely to occur, which means that D–D plasmas produce dramatically fewer fusion neutrons. In a power plant, these neutrons are the goal, but in an experiment they complicate operations enormously. The fusion neutrons hit the components surrounding the plasma, making everything sufficiently radioactive that scientists can no longer go inside the machine. As discussed last chapter, this necessitates performing all maintenance remotely via complicated robotics. In a power plant this is difficult, but in an experiment it's even more of a problem. We want fusion experiments to be experimental. We want to be able to install new components on a whim and fiddle with unruly diagnostics. Remote maintenance

---

[28] The cafeteria at the JET site stopped serving beer just before tritium arrived on site. Coincidence?

makes the machine less flexible — or at least it makes flexibility more time-consuming.

Hence, until the 1990s, no large tokamak had ever created a D–T plasma, nor generated substantial fusion energy. This, of course, was unsatisfactory. Without substantial fusion energy, you can't study plasma self-heating, i.e. how the high energy helium nuclei produced by fusion slow down and transfer their energy to the plasma. Additionally, no one was really sure how similarly pure deuterium plasmas and deuterium–tritium plasmas would behave. From the plasma's perspective, the only difference is that 50% of the nuclei now have 50% more mass. The general consensus was that the difference probably wasn't too important ... probably.

Thus, to inform the on-going ITER design (which we will discuss next chapter), the global fusion community set out to experimentally investigate these issues. In 1991, JET performed a small set of tritium experiments and generated the first megawatt of controlled fusion power (see Figure 6.12). However, these experiments used only a limited amount of tritium in order to minimize the induced radioactivity of the device. Soon thereafter, another opportunity presented itself. The American flagship tokamak, TFTR,

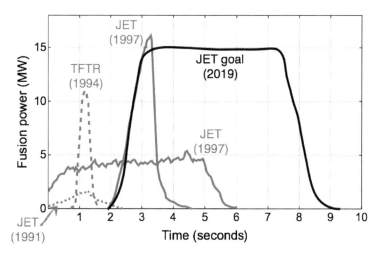

Figure 6.12: The best results from past D–T experiments (gray) as well as a goal for the upcoming JET D–T experiments (black) [28, 29].

was slated for shutdown, so it went out with style. It concluded its operation with an extensive set of D–T experiments — one of which generated over 10 megawatts of power. Finally, in 1997, JET bit the bullet and installed a sophisticated remote maintenance system. This enabled the most extensive and ambitious set of D–T experiments. Leveraging its "D"-shaped plasma cross-section (TFTR was circular), JET generated over 16 MW of fusion power and reached $Q = 0.7$. Moreover, JET confirmed some encouraging results from TFTR — adding tritium to the plasma actually improves its performance![29]

Since then, there have been no further D–T experiments, which is why there have been no new records for fusion power. However, there is some excitement on the horizon. JET plans to perform a third set of D–T experiments in 2019. Unfortunately, it doesn't look like JET will achieve $Q = 1$ this time around either. In order to make it more useful for predicting behavior in ITER, some modifications were made that somewhat reduced its performance.[30] That being said, they do have some ambitious plans (see Figure 6.12).

6.9   2000s: Supercomputers

Much of our information regarding what happens in plasmas does not come from diagnostics. Rather, advances in the power of computers over the past several decades have revolutionized the way we understand fusion devices. Before about 1980, pencil and paper math was the most powerful technique available, so physicists were forced to use simplified models like MHD and devise clever mathematical techniques to approximate the plasma's behavior. The limits of what could be done proved quite restrictive. To circumvent this, physicists

---

[29] Specifically, the external heating power needed to get into H-mode was found to decrease with the average mass of the nuclei. This is called the *isotope effect* and it is an active area of research.

[30] Because it possesses a larger plasma, ITER needs a more robust first wall and divertor (made from beryllium and tungsten, respectively). In order to gain operating experience with these materials, the JET components were changed, which hurt performance somewhat.

now employ computers to solve formerly impossible problems and arrive at a more complete understanding of the plasma.

There have been two breakthroughs that enabled the recent successes of computational plasma physics. The first is obvious. Anyone with a laptop has experience with it. In the last few decades, the computational power of computers has exploded. This rapid rate of progress, embodied by Moore's law, has held since the early 1970s. Fusion is in no way responsible for this, but is particularly well-suited to take advantage of it. Problems in plasma physics are fairly special in that they are extremely parallelizable. By this, we mean that problems can often be split into parts, which can be solved in isolation and then assembled to determine the overall solution. So if you use, say, four CPUs in your computer together, you could solve a plasma physics problem about four times faster than if you had only used one CPU. For this reason, plasma simulations are run on some of the world's biggest supercomputers, which can have hundreds of thousands of CPUs each. Every few years, a new one is built that can perform calculations significantly faster, enabling even better simulations of plasma behavior.

The second breakthrough came from within the fusion community. Plasma physicists developed innovative ways to reduce the computational time of simulations, without much loss in accuracy. Understanding the relevant physics at a deep level allows computer calculations to be streamlined significantly, enabling the same computer to perform a simulation more quickly. This is possible because, to extract useful information, you don't actually have to include every single physical detail.

Naively, simulating plasma appears easy because we already know the fundamental equations that govern its behavior.[31] We know how each particle moves through electric and magnetic fields (see Figures 4.1 and 4.3) and we know how each particle generates electric and magnetic fields (see Figures 4.2 and 4.4). Hence,

---

[31] Every particle moves according to the Lorentz force law and generates electric and magnetic fields through Maxwell's equations. That's all there is to it.

the most realistic approach would be to model all the particles individually. In theory, this brute-force method works perfectly. In practice, it's completely useless. The problem is there are a tremendous number of particles in the plasma (roughly 10^{23}) that gyrate around the field lines very quickly (roughly once every 10^{-12} seconds). Explicitly modeling this behavior for just a fraction of the energy confinement time turns out to require 10^{31} years on today's most powerful supercomputer! So, what now? What can we do to enable current computers to give helpful information?

There are a number of simplified models of the plasma that can enable simulations to answer different questions to various degrees of accuracy. As an illustrative example, we will explain *gyrokinetics*, which is a high-fidelity model aimed at simulating plasma turbulence and calculating energy transport[32] (see Figure 6.13). Its mathematical development began in the 1960s, but it took until the early 2000s until computers were sufficiently powerful to make use of it. While gyrokinetic computer codes are challenging to develop, run, and interpret, they have been overwhelmingly successful.[33] Gyrokinetics has been used to estimate the energy confinement time, evaluate proposed experiments before they are built, and illuminate some of the physical processes that underlie H-mode.

The first step in producing a tractable model isn't specific to gyrokinetics. The biggest shortcut we can take is to significantly reduce the number of particles we simulate. While a tokamak can contain 10^{23} particles, they behave so similarly to one another that we don't need to model them all to get an accurate answer. In practice, we can get by with just 10^9 particles, which dramatically reduces the computational time from around 10^{31} years on the world's most powerful supercomputer to just a 1,000 years. However, this isn't quite enough.

---

[32] If you are interested in a technical description of gyrokinetics, Ref. [30] is a good place to start.
[33] Gyrokinetic codes are hundreds of thousands of lines long and often written in Fortran.

Figure 6.13: The turbulent fluctuations in plasma density as calculated by a gyrokinetic simulation. Black indicates a slight accumulation of ions and white indicates a slight accumulation of electrons.

To enable practical simulations, gyrokinetics relies on some complex trickery. Specifically, computational physicists exploit what is called *scale separation* to remove unnecessary physics in the simulations. Scale separation is the idea that behavior at different scales in space and time can be (to some degree) decoupled and treated independently. However, doing this without destroying the accuracy of the simulation requires deep physical insight into the problem and a sophisticated mathematical framework. Accomplishing this for gyrokinetics was a long, drawn-out effort that took the fusion community roughly 30 years. The defining insight for gyrokinetics was noticing that particles oscillate around magnetic field lines thousands of times faster than any other behavior. This allows the gyration to be separated out and solved on its own. In fact, the solution is simple: particles make circles in the plane perpendicular to the field lines. Hence, physicists realized that by simulating *rings of charge following magnetic field lines* instead of *charged particles oscillating around field lines*, they could avoid having to simulate the gyration at all (see Figure 6.14).[34] The primary benefit of

---

[34] While this may seem simple, it took 30 years to develop for a reason. There are all sorts of complications. We must distinguish between motion along the field

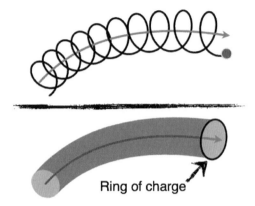

Figure 6.14: The gyration of a charged particle around a magnetic field line (top) and the equivalent ring of charge that is modeled in gyrokinetics (bottom).

gyrokinetics is that, instead of having to explicitly model behavior that takes 10^{-12} seconds (i.e. the particle gyration), the fastest behavior you must model takes 10^{-6} seconds (i.e. the lifetime of turbulent eddies). Using this factor of a million, our simulations now only require the entirety of a supercomputer for about a day.

These sorts of simulations are now feasible and have been performed. However, commanding an entire supercomputer, which is a multi-million dollar machine, for a full day is costly and stressful. Imagine that, after the simulation finishes, you discover that you made a mistake! For that reason, we usually only simulate a fraction of the device, generally the plasma immediately surrounding a few magnetic field lines. This reduces the run time of the simulation a bit further, making it a lot more practical. You can request a thousand CPUs on a supercomputer and get the results of your simulation in less than a day. These cheaper gyrokinetic simulations enable

---

line and motion across it (while still properly treating particle drifts). Particles have different radii of gyration, which change as they move around the device. We define the location of the rings of charge by their centers, but then must adjust for the fact that it experiences the electromagnetic fields along the ring itself (not at its center).

researchers to develop their intuition, try things out on a whim, and make mistakes. This is how cutting-edge science is actually done.

Importantly, gyrokinetics is just part of the picture. There are other theoretical models that are useful for different purposes. For example, if you are designing a heating system and want to know where electromagnetic waves are absorbed by the plasma, then gyrokinetics is fairly useless. This is because gyrokinetics doesn't model the particle's gyration, which is important physics for wave heating. Remember, absorption typically relies on matching the oscillation of the wave with the oscillation of a particle around its field line. Therefore, the community has developed *Fokker–Planck* simulations, which directly evolve individual particles similarly to the brute-force method mentioned above. Of course, the downside is that they require so much computing power that they can't simulate the plasma for long enough to observe turbulence.

On the other side of things is MHD, which we already discussed in Chapter 4. It throws away even more physics than gyrokinetics. It doesn't track individual particles, but instead models the plasma as an electrically charged fluid. Still, it is useful for predicting the overall bulk motion of the plasma, just not for fine-scale details like turbulence. In the early days of fusion, MHD was essential because it was simple, easy to use, and told experimentalists the conditions under which the plasma would go wildly unstable. Before MHD stability was understood, the plasma in experiments didn't exist for long enough to develop turbulence.

With recent advances in computational plasma physics, simulations are finally approaching the point where they can predict important aspects of experiments (even including turbulence).[35] Moreover, in the coming decades, increases in computing power seem certain to improve matters further. Simulations will become more accessible and more realistic. Finally, while the mathematical work needed to simulate an isolated magnetic fusion plasma is fairly

---

[35] As mentioned in Chapter 4, computation has also been crucial in optimizing stellarators.

mature, important challenges remain elsewhere. Research on the edge of the plasma and how it interacts with solid material and neutral gas is only just beginning. Understanding this behavior is critical to answer questions like "Will the power exhausted by the plasma melt the divertor?" Moreover, the plasma edge appears to play an important, but unknown role in enabling H-mode. Perhaps better simulations could reveal strategies to reduce turbulence throughout the entire device.

Chapter 7

The Present: ITER

ITER is a huge tokamak and one of the most ambitious science experiments ever. It is expected to be the first device in human history to confine a plasma that generates more fusion power than the heating power needed to sustain it. ITER, alongside scientific megaprojects like the International Space Station and the Large Hadron Collider, help to define humanity and our capacity for international collaboration. While you may not realize this, chances are high that ITER is being built in your name. This is because it is the biggest scientific collaboration ever, with 35 countries (representing over half the world's population) contributing time, money, and components. So, if you live in the European Union,[1] the USA, China, India, Japan, South Korea, or Russia, ITER will be part of your legacy to future generations.

While ITER started out as an acronym for the International Thermonuclear Experimental Reactor, it was quickly realized that having the words "experimental" and "thermonuclear" so close to one another wasn't great for public relations. Now, ITER simply signifies the Latin word for "the way."[2] The assembly of ITER has already begun in southern France out of components that are being constructed all over the world. As of 2017, it stands approximately halfway completed. After it starts operating around 2025, *ITER aims to demonstrate the scientific and technological viability of*

---

[1] Or Switzerland.
[2] It's cleaner, don't you think?

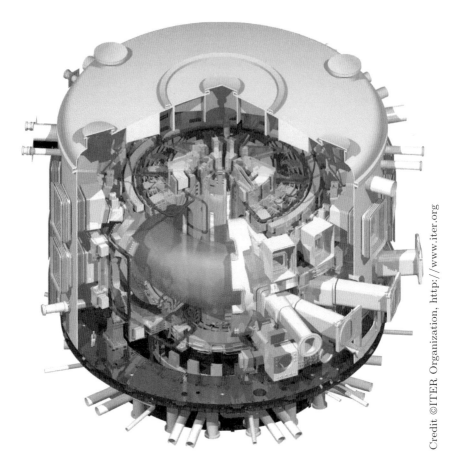

Figure 7.1: The ITER tokamak. Note how big the device is compared to the person in the bottom left corner.

fusion energy. While the goals and ideals of ITER are laudable, its implementation has been ... complicated. In this chapter, we will discuss ITER's goals, how it will achieve them, its cost, and where it might lead.

7.1 ITER's Goals

ITER aims to be the link that connects existing experimental devices to the first demonstration power plant. ITER will produce no electricity and is both a physics and engineering endeavor.

It will test how well we understand magnetic confinement fusion and illuminate the behavior of large high-performance plasmas. If ITER is successful, a path to a demonstration tokamak power plant will be, for the most part, clear. The important physics would be largely settled, making power plant design primarily engineering and economics. However, if ITER fails, it may indicate that the scientific and technological basis of the tokamak is not sound. Thus, the stakes for ITER are high.

To serve this purpose, ITER is designed to attain the following goals:

(1) **Achieve a plasma power multiplication factor of Q > 10 in 5 minute long pulses.** ITER is designed to create deuterium–tritium plasmas that significantly exceed breakeven (i.e. $Q = 1$). A controlled plasma has never generated more fusion power than the external heating power needed to sustain it — let alone by a factor of 10! In order to do this, ITER must generate 500 MW of fusion power — thirty times more than any tokamak before it. Additionally, this plasma will be sustained for over 5 minutes — almost a hundred times longer than any other D–T plasma.[3] Still, these pulses are short enough that ITER can rely entirely on inductive current drive to sustain the plasma current. Remember, in Chapter 5 we learned that inductive current drive, which uses the central solenoid, is inherently pulsed and cannot be sustained indefinitely. Non-inductive current drive (e.g. neutral beams, electromagnetic waves, and/or the bootstrap current) is more difficult, but can be sustained *forever*.

(2) **Achieve a plasma power multiplication factor of Q > 5 in steady-state discharges using only non-inductive current drive.** Future power plants will not be able to rely on inductive current drive, except for a short time during start-up. Thus, ITER is also designed to demonstrate an impressive Q with only non-inductive current drive. This is important because steady-state operation can assess the importance of

---

[3] JET achieved $Q = 0.2$ for a few seconds.

slowly-evolving interactions between the plasma and the solid materials surrounding it. For example, the divertor will gradually absorb deuterium and tritium (changing its material properties). Additionally, sustained exposure to the plasma can cause small flakes of solid material to chip off from the first wall. Lastly, steady-state operation is needed to produce a large number of neutrons in order to test their effects on materials (e.g. tritium breeding).

(3) **Not preclude the possibility of achieving controlled ignition (i.e. Q = ∞).** While ITER is not designed to achieve ignition, if things go well, it still could. The fusion technology surrounding the plasma has been built to handle an ignited plasma. However, the best predictions for the performance of ITER indicate that achieving the confinement time required for ignition is uncertain.[4] That being said, by the time ITER shuts down decades from now, the community may have discovered new ways of boosting plasma performance.

(4) **Achieve a burning plasma.** This is a direct consequence of goal (1), but we distinguish it to stress its importance. A *burning plasma* is defined to be a plasma that is mostly self-heated. One-fifth of all fusion power is carried by the helium-4 fusion product (as opposed to the neutron). Since the helium-4 has an electric charge, it is confined by the magnetic field and deposits its energy in the plasma as heat. So when $Q = 10$, the plasma's self-heating will be 2 times larger than the external heating power. Significant self-heating has never been achieved before, but will be the dominant source of heating in a power plant. Thus, it is important to explore and exploit it. So, how are burning plasmas different from your garden-variety externally heated plasma? The largest difference is the abundance of helium-4 fusion products. These guys are born with such high energies that they can trigger special types of plasma instabilities. Moreover, we must

---

[4]The triple product $nT\tau_E$ (i.e. density times temperature times energy confinement time) required for ignition is 50% larger than the triple product required for $Q = 10$.

ensure that the helium-4 is confined long enough for them to transfer their energy to the fuel, but not so long that they build up and obstruct more fusion. Thankfully, a large amount of theoretical and computational research has focused on how to manage burning plasmas. The exploration of burning plasmas is one of the most significant and highly anticipated plasma physics questions that ITER will address.

(5) **Demonstrate the integrated operation of essential fusion technologies.** For its construction, ITER required a plethora of next-generation technologies for remote maintenance, materials, heating, diagnostics, cryogenics, magnets, and plasma control. Once it begins operations, it will test these components in a burning plasma environment fairly similar to that expected in a power plant. All of these technologies will be vital to the successful operation of future fusion power plants.

(6) **Test tritium breeding schemes.** In its later years, a number of tritium breeding test modules will be installed on ITER. These will try out different concepts for ensuring a future power plant can achieve tritium self-sufficiency. Since ITER will be the first device capable of producing a large, continuous flux of fusion neutrons, tritium breeding has never been experimentally tested before.

(7) **Demonstrate the safety characteristics of a fusion device.** ITER will show the world that a fusion power plant will have negligible negative environmental consequences.

As a caveat, we should remind you that these Q values do not include the power required to run all of the ancillary power plant systems. It is simply the *plasma* power multiplication factor. Even though ITER's Q will be much greater than one, the engineering power plant multiplication factor, $Q_{\text{eng}} = P_{\text{output}}/P_{\text{input}}$, will be zero. While some may see this as a failing, it misunderstands the purpose of ITER. ITER is about demonstrating the science and technology of a fusion reactor. It is not optimized for net energy production and will not deliver any electricity to the grid. That will be a task for the device after ITER: a demonstration power plant.

Nevertheless, ITER's goals are stunning. One of the most important things to notice is that they are not simply incremental improvements. ITER will illuminate several unexplored physical mechanisms for the first time — it will host burning plasmas, trial tritium breeding facilities, and test components in a fusion neutron environment. Investigating any one of these would advance the field of fusion substantially. Yet, ITER aims to achieve them all in a single integrated device, making it even more attractive. The obvious questions then becomes "How does ITER intend to do this?" and more importantly "Is it expensive?" These questions will be the topics of the next two sections.

7.2 ITER's Strategy

While we've just seen that ITER has lofty goals, it employs a fairly simple strategy to reach for them. The advantage ITER has over existing tokamaks is size. ITER's performance is expected to be so awesome primarily due to how large it is. It dwarfs even the largest of present-day fusion experiments (see Figure 7.2).

	JET	ITER (projected)
Torus radius	3.0 meters	6.2 meters
Cross-sectional radius	0.9 meters	2.0 meters
Plasma volume	100 cubic meters	1,000 cubic meters
Power multiplication factor, Q	0.7	10
Fusion power	16 MW	500 MW
Total heating	26 MW	50 MW+
Longest H-mode plasma	20 seconds	5 minutes+
Plasma current	7 MA	17 MA

Figure 7.2: A comparison of ITER's projected parameters against JET, the largest existing fusion experiment.

Here are the numbers. The ITER site, including all of the supporting buildings, will cover almost a full square mile (i.e. 2 square kilometers). The tokamak itself will comprise over 18,000 tons of steel and machinery. The external coils used to generate the confining magnetic fields will contain 100,000 kilometers of superconducting niobium–tin wire. This magnetic field will permeate a huge donut-shaped vacuum vessel, which has an overall diameter of 20 meters and height of 11 meters. All of this to enable 1,000 cubic meters of plasma to reach a temperature of hundreds of millions of degrees. Indeed, ITER is a monumental undertaking.

So why does ITER have to be so large? We'll learn a more sophisticated answer in the next chapter, but, basically, it is to have a long enough energy confinement time to attain $Q = 10$. The confinement time is limited by plasma turbulence, which takes energy from the center and gradually moves it out of the device. This gives us two avenues to improve confinement: reduce turbulence to slow the transport of energy or increase the distance the energy must travel to escape (see Figure 7.3). While fusion research has

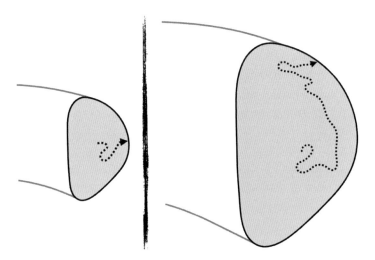

Figure 7.3: The poloidal projection of an ion's turbulent random walk in JET (left) versus ITER (right). We see that, even if energy moves outwards at the same speed in both devices, it will take longer to escape from ITER.

discovered several strategies to reduce turbulence (e.g. H-mode), they are already being employed to the best of our abilities. Thus, until further breakthroughs are made, our only avenue for progress is to make bigger devices.

Intuitively, this makes sense. Bigger should be better because the fusion power increases with the plasma volume, while the power leakage increases with the surface area. In other words, power is produced all throughout the plasma, while power can only escape through the surface. Thus, since volume is proportional to the radius *cubed* and surface area is proportional to the radius *squared*, we would expect the energy confinement time to be proportional to the plasma radius itself (i.e. doubling the radius would double the confinement time). In reality, things are considerably more complex, but when the math and physics are accounted for to the best of our knowledge, it appears that ITER's size will be sufficient to satisfy its ambitions. So that, in a nutshell, is why ITER is so darn big: if you make your device bigger, it takes longer for things to get out. Unfortunately, we don't get this better confinement for free. Not only does ITER's size make it more expensive, but it makes its fusion technology more challenging — most notably, the plasma heating systems, the divertor, and the first wall.

7.2.1 *Heating systems*

The first concern is how to heat the plasma. ITER is so big and hot that, even though most of the heating power will come from the plasma itself, the amount of external heating in absolute terms is still very large. The best estimates indicate that 50 MW of external heating will be necessary to achieve the temperatures for fusion. Even if ITER were to ignite, powerful heating systems would still be needed to reach the ignited state. While ITER is not pioneering any brand new heating schemes, many of the existing techniques are challenging to implement because of ITER's scale and neutron production.

The most powerful component of ITER's heating system will be its two neutral beam injectors. Both will be physically massive — with lengths similar to the diameter of the entire reactor (see

The Present: ITER

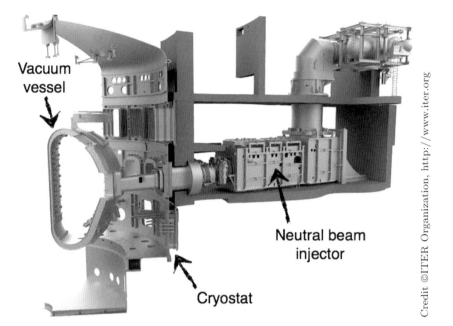

Figure 7.4: One of ITER's two neutral beam injectors. The beam passes through ports in the cryostat (which keeps everything within it at cryogenic temperatures) and the vacuum vessel.

Figure 7.4). Each of the injectors will accelerate deuterium atoms in order to deposit 17 MW of heating power into the plasma. However, what's tricky isn't achieving the *total power* carried by the beam, but rather the *energy of the individual particles* that comprise the beam. ITER is so large and dense that, unless the individual particles are moving really fast, they all will slow down and stop in the edge of the plasma. Since we want to heat the center of the device, each particle must be given an extraordinary amount of energy: 1 MeV. Unfortunately, at these energies, it is difficult to neutralize an accelerated particle beam. Remember, typically a neutral beam injector works by taking neutral atoms, removing an electron from each of them, accelerating them, and then adding electrons back again. However, for ITER, the beams will be moving so fast that the final step of adding the electrons back again becomes prohibitively difficult.⁵ Thus, ITER needed to pioneer a new bit of neutral beam technology, specifically what is called *negative ion acceleration*. The idea is to add an extra electron to a neutral atom and then remove it after acceleration. While the first step of adding the extra electron to neutral atoms is tricky, it is easy to knock them off after acceleration, even at high energies. A huge amount of work has gone into developing negative ion acceleration technology and, if it works as advertised, it will be one of the success stories of ITER.

The rest of ITER's heating power will be provided by two systems using electromagnetic waves: electron cyclotron heating and ion cyclotron heating (see page 150 for a reminder). Each system uses antennae to blast 20 MW of power into the plasma, where it is absorbed by either the electrons or the ions. You may have noticed that all of these heating schemes — 33 MW of neutral beams, 20 MW of electron cyclotron waves, and 20 MW of ion cyclotron waves — add up to more than 50 MW. The reason for this is that, during the first 10 years of ITER's operation, it will use normal hydrogen and/or helium fuel, rather than deuterium–tritium.⁶ Thus, the plasma will produce little fusion power, so the external systems must compensate for the lack of self-heating.

7.2.2 Divertor

Another concern, the divertor and power exhaust system, is often cited as ITER's biggest engineering challenge. With its 1,000 cubic meter D–T plasma, ITER will produce a lot of fusion power — 500 MW. Obviously, this is good as it is the ultimate goal, but it can still be hard to manage. ITER's divertor plate is estimated to have

---

⁵Specifically, the size of the charge exchange cross-section (which adds electrons) decreases relative to ionization cross-section (which removes electrons) as the particle energy increases.

⁶Remember that current experiments use pure deuterium fuel (instead of D–T) to reduce the production of fusion neutrons and prevent the surrounding components from becoming radioactive. ITER will be so good at fusion that even creating pure deuterium plasmas would generate too many neutrons.

as much as 20 megawatts of power deposited on each square meter of its surface. That's enough to power 40,000 British households *per* square meter.[7] In fact, ITER's divertor plate will bear the highest steady-state heat flux of any solid material on Earth. Tackling this immense challenge requires some intense engineering. Multiple sophisticated techniques are being employed in concert to ensure that the plate won't melt. That said, the design itself is fairly simple (see Figure 7.5). It is a solid piece of tungsten that has coolant channels running through it. Water is circulated through these channels at high speeds in order to carry away the heat. Tungsten was chosen for its extremely high melting point ($3422°C$), high thermal conductivity (meaning that heat moves through it quickly), and high durability. Moreover, tungsten does not react chemically with the deuterium and tritium that will be bombarding it.[8]

One technique that will be used to reduce the heat flux is to tilt the divertor plate such that the magentic field lines hit the solid material surface at grazing incidence (see Figure 7.6). By simple geometry, this increases the area over which the power is deposited. A second techniques uses an extra poloidal field coil to make the poloidal magnetic field flare out near the divertor. Since the field lines

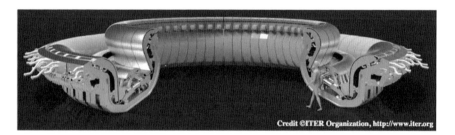

Figure 7.5: A person standing in front of the ITER divertor plate, which comprises 54 individual cassettes that fit together to form a full toroidal circle in the bottom of the vacuum vessel.

---

[7]But only 14,000 American homes.
[8]Tungsten is basically the Rocky Balboa of materials (though I do imagine that Rocky would react if bombarded with tritium).

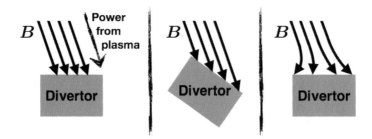

Figure 7.6: The power leaving the plasma follows the magnetic field lines (black) until it hits the solid divertor plate (gray). The usual setup (left) can be altered to reduce the power per unit area by tilting the divertor (center) and/or flaring the magnetic field lines (right).

spread apart, particles traveling along them will also be spread out. These (and several other strategies) should enable ITER to handle the greatest steady-state heat fluxes anywhere on Earth.

7.2.3 First wall

Just beyond the edge of the plasma sits ITER's first wall. It will be mounted to the inside of the world's largest vacuum vessel, which will weigh nearly as much as the Eiffel Tower (see a single toroidal section of it in Figure 7.4). During normal operation, the first wall will experience considerably less power than the divertor, but it can still take a beating. One reason is ELMs, which are the periodic bursts of energy from the edge of an H-mode plasma that we discussed in Chapter 6. In a device as big as ITER, continuous exposure to strong ELMs could erode away the material of the wall in just a couple of months. For this reason, ITER will use both ELM suppression techniques discussed in the Tech Box on page 195: pacing with small pellets and creating stochastic magnetic fields in the edge. More concerning for the first wall is the possibility of a disruption. The plasma is so large that a full-scale disruption would have more than enough thermal energy to melt the surface of the wall. Several of these disruptions could put the tokamak out of action, requiring months to replace the damaged components. So, what will ITER do to avoid disruptions and reduce their severity?

Well, first off, ITER will undoubtedly be operated very conservatively, especially during its early days. This means running the machine at low density, temperature, and electric current so that the disruptions are much weaker. As operators gain experience with the machine and better understand the disruption risks, more aggressive operation can be pursued. However, regardless of how careful they are, strong disruptions will almost certainly occur. For that reason, the disruption mitigation system must be made very reliable. ITER will use shattered pellet injection (as described in the Tech Box on page 155) in order to radiate away the plasma's energy before it touches the first wall. Though similar systems have been tested to some extent on existing tokamaks, it remains an especially high-stakes design challenge. After D–T operation starts, any damage will have to be repaired remotely using a robotic system, which is another design challenge in and of itself.

Because the thermal and mechanical forces on the first wall will be so large (and maintenance will be so awkward), it is important for ITER to have the most robust wall possible. Thus, it would make sense to use a tough material like steel. However, because the wall is so large and entirely surrounds the plasma, there are other considerations. Most notably, interactions between the wall and the plasma can be very important. When high energy particles from the plasma hit the wall, they can knock atoms out that then travel into the plasma. This means you have to be careful about your choice of wall material — whatever you choose will undoubtedly end up in the plasma!

So what happens when a particle from the wall gets into the plasma? One of the main concerns is electromagnetic radiation. Since deuterium and tritium atoms only have a single electron, the thermal energy of the plasma is more than sufficient to remove it. However, heavy atoms (like the iron in steel) have a lot more electrons, many of which are very tightly bound to the nucleus. The plasma isn't hot enough to remove all these electrons. Instead, the plasma continually excites them to higher atomic energy levels and, when they drop back down, they release electromagnetic radiation. These heavy atoms can eject energy from the plasma even faster than turbulence, ruining the confinement time. So despite the fact that steel could handle disruptions very well, you would not want to make the entire first wall out of it.

The obvious solution is to make the wall out of a material that only has atoms with a low number of protons. When these atoms get sputtered into the plasma, all their electrons get removed, so they radiate little. For this reason, ITER's first wall will be made of beryllium, which only has four electrons. While beryllium has a fairly low melting point (1287°C instead of tungsten's 3422°C), it enables better plasma performance. Another essential feature of beryllium is that, because of its chemical properties, it doesn't absorb tritium. Other materials like carbon would be ideal for the wall, except that they soak up tritium like a sponge. Since tritium is radioactive and costs $30,000 per gram, this more-or-less disqualifies them as a first wall material.

Finally, one last bonus of beryllium is that it multiplies neutrons (as discussed on page 162). When neutrons from D–T fusion pass through the first wall, some will strike beryllium nuclei and induce the beryllium neutron multiplication reaction. This will increase the total number of neutrons produced, which is good. We want lots of neutrons to test out different tritium breeding strategies.

7.3 ITER's Schedule and Cost

ITER has had a long and tumultuous history, in which cost has played a key role (see Figure 7.7). The first step towards a huge internationally funded tokamak occurred in 1973, when Soviet General Secretary Leonid Brezhnev and US president Richard Nixon met in Washington D.C. Optimism about fusion was high. The worldwide oil crisis was in full swing and, as a result, funding for fusion had never been better (and hasn't been as good since). It was at this time that the present-day fleet of large national tokamaks was designed and built (e.g. JET in the EU, TFTR in the US, JT-60 in Japan, and T-15 in the USSR). During their meeting, Brezhnev and Nixon agreed that an international collaboration in fusion energy would be a productive way to ease Cold War tensions. Soon afterwards, they,

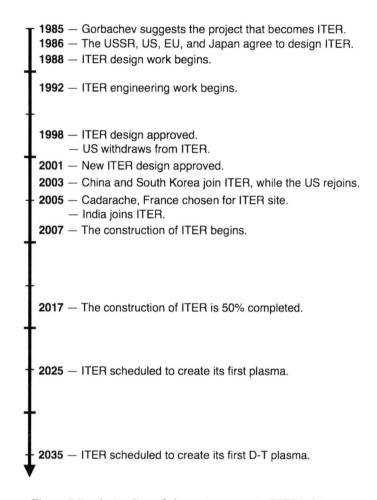

Figure 7.7: A timeline of the main events in ITER's history.

along with the EU and Japan, committed to explore the potential of an internationally funded tokamak. The project became known as INTOR, the INternational TOkamak Reactor. Between 1978 and 1988, hundreds of scientists and engineers would meet in INTOR workshops several times a year. At these meetings, the different research groups would be assigned "homework" to take back to their institutions and (hopefully) complete before the next meeting. While the project never reached the engineering design phase, it scoped out the potential for such a device and was well received by the next Soviet General Secretary, Mikhail Gorbachev.

At the 1985 Geneva Summit, Gorbachev proposed the ITER project to US president Ronald Reagan as a continuation of INTOR. While ITER was similar to INTOR, it had a significantly different organizational structure. The INTOR organization, which was basically a loose confederation of scientists, was too decentralized and weak to effectively manage an actual construction project. The ITER organization was designed to be much more powerful, but (as we will soon see) would still struggle to manage such a large and unwieldy endeavor. The largest international scientific collaboration ever would turn out to be as much a political experiment as a scientific one.

In 1986, the United States, the Soviet Union, the European Union, and Japan officially signed the ITER agreement, with the aim of designing and building the next generation tokamak experiment. The original goals of ITER were even more ambitious than they are now. ITER would ignite, not merely achieve $Q \geq 10$! Given that, at this time, no large device had even operated with tritium, going to a fully ignited device was downright daring. It's not that it was impossible, but it was a striking plan. Design work commenced, resulting in a "final" device 12 years later. The tokamak was a behemoth. With a toroidal radius of 8.1 meters, a plasma cross-sectional radius of 2.8 meters, and a thermal fusion power of 1,500 MW, it was roughly three times bigger than the device currently being built today. Moreover, the device was estimated to cost $11 billion (adjusted to 2017 dollars), which everyone agreed was almost certainly optimistic.

The US balked at such a high price ... and promptly left.

At this point, ITER seemed pretty dead. But working with a small staff and a small budget, the remaining participants substantially redesigned the experiment in just three years. By watering down the performance goals somewhat, they designed a smaller device that they could credibly advertise as less expensive. This design is what is currently being built. Instead of igniting, ITER would achieve $Q > 10$. Instead of producing 1,500 MW of fusion power, it would make 500 MW. The total construction costs were estimated to be a paltry $6 billion. The redesign worked. By shifting the balance between performance and cost, the ITER team was able to entice China, South Korea, and India to join, and the US to rejoin. And so ITER was reborn.

The next step, choosing the location to build the machine, went surprisingly smoothly.[9] Four sites were proposed: Rokkasho in Japan, Cadarache in France, Vandellos in Spain, and Clarington in Canada. In 2005, the negotiations finished without even requiring a major international emergency to conclude them. France won and got ITER, but had to pay a steep price. As a buy-out, the EU gave Japan 20% of the project's scientific positions (including the director-general role), money to help upgrade their national tokamak JT-60U, and a supercomputer. Moreover, the EU agreed to pay Japan to make some of the components that were allocated to the EU, thereby helping to develop Japan's fusion manufacturing capabilities. Spain, in turn, got the financial headquarters that controls all of the EU's spending on ITER. Canada got nothing and left the project.[10] It was also at this point that the relative contributions of the different parties were determined. Because the physical presence of ITER carries significant economic and scientific benefits, the EU agreed to pay 45% of the cost. The other six member states each agreed to pay an equal share of the remainder, which came out to 9%.

With the design finished and the site chosen, construction began in 2007. At this time, the plan was for ITER's first plasma to be in 2018. Unfortunately, as construction progressed, the schedule began to slip and costs skyrocketed. As of 2017, ITER has the first plasma scheduled for December 2025 and the total construction costs to be around $25 billion. What happened?! The reasons are, of course, complicated and unclear. While substantial delays and cost

---

[9]When we say "surprisingly smoothly," we mean that choosing a site didn't appear to actively delay the entire project substantially. It was still incredibly contentious as you can imagine.
[10]Oh Canada...

overruns certainly aren't unusual for a large scientific project, ITER's difficulties go beyond that.

The first issue was that the $6 billion figure thrown around in 2001 explicitly did *not* account for real-world manufacturing expenses, let alone the intricacies of an international collaboration. Moreover, this figure was estimated *before* the design was finished. So it appears that $6 billion was an idealized, preliminary estimate that was then interpreted as a prediction of the ITER price tag.

Another sink of time and money is the way that components were allocated. Since some components represent new technologies and are therefore more desirable to build, the seven member states fought over them. Every country wanted to develop their own domestic fusion manufacturing base by building the exciting, fusion-specific components (e.g. the magnets, vacuum vessel, heating systems). Hence, the allocation of manufacturing tasks quickly became a contentious political process, rather than one of engineering shrewdness. For example, ITER's vacuum vessel was designed to be constructed on-site in Cadarache from nine separate toroidal sections that are largely identical. Instead of one country making all nine, a compromise was necessary: the Europeans got seven and the South Koreans got two. This was troubling because all nine pieces must fit together flawlessly and have exceedingly tight tolerances. Therefore, it caused an uproar when word arrived at the ITER site that the Europeans were designing their sections to be bolted together, while the South Koreans expected theirs to be welded. From a project management standpoint, this boggles the mind.

Perhaps the most important consideration is the decentralized organizational structure of the project. Contrary to practically all other international scientific collaborations, ITER has no central budget and no member state has full control of the project. Moreover, the central ITER organization is a relatively small workforce of less than a thousand people, who act mainly as coordinators. Most of the work for ITER is handled by the different *domestic agencies*, which are organizations that each member country has set up to handle the procurement of its own components. These domestic agencies are responsible for supplying a lot of hardware, but relatively little cash. Thus, ITER is primarily constructed around the world. Then, the finished components from each country are sent to France and assembled on site. These domestic agencies have free reign to manage their own budgets and are only tasked with providing their assigned components in a "timely" fashion. The central ITER organization has little leverage over the domestic agencies and even less over the national governments that fund them. Hence, it is more-or-less impossible for ITER to put together a rigorous schedule or budget — both strongly depend on the national governments funding their domestic agencies as intended, which sometimes does not happen (e.g. in 2017, the US contributed just $50 million instead of the $200 million that was expected). The most incredible thing is that, even after ITER is fully built, we won't precisely know the construction costs — the domestic agencies have no obligation to disclose their spending.

In early 2014, a scathing assessment of ITER's management was leaked to the press.[11] The report highlighted the inadequacy of the original ITER design, the distributed manufacturing of common components, a lack of personnel with experience in project management, and poor communication between the ITER central organization and the domestic agencies. It recommended that the director-general be replaced, the management be significantly restructured, and a transparent, realistic construction timeline be developed. Soon thereafter, a new director-general, Bernard Bigot, took the reins and December 2025 was identified to be "the earliest possible technically achievable date" for the first plasma.[12] The results of this shake up are still unclear, although early reports are positive. However, no matter how capable the ITER management becomes, if the member nations do not fund the project as promised, the schedule will continue to slip.

Despite all of these challenges, ITER is going ahead (see Figure 7.8). The path has been winding and arduous — and yet,

---

[11] For a 2014 perspective on the ITER project, we recommend *A Star in a Bottle*, an article in the New Yorker by Raffi Khatchadourian.

[12] The calculation of this date includes no contingency, so it is optimistic.

Figure 7.8: An aerial view of the ITER site in July 2017. No, the buildings are not photo-shopped.

slowly but surely, the world's most complex scientific apparatus is coming together. Buildings are going up. Equipment from all over the world is converging on Cadarache. All the superconducting cable has been produced and some of it has been used to build ITER's first toroidal field magnet (see Figure 7.9). As of late 2017, ITER stands roughly half completed and we are hopeful that, soon after 2025, the first plasma will fire up. In the years that follow, more components will be added to the machine, expanding its capabilities. Operators will gradually build up their experience and learn how best to pilot it. This will all culminate circa 2035, when D–T operation will commence. Finally, the full power of ITER will be mustered in an attempt to unleash the energy source of the future.

While ITER has had a profound impact on every country's fusion program, its cost has been felt especially strongly in the United States. America's relationship with ITER has been particularly complicated, so it is not surprising that it has seemed ill-prepared to balance ITER with its own domestic fusion program. On one hand, the US Congress has been allocating just a fraction of the money that ITER has been promised, citing "concerns about the cost and

Figure 7.9: The first ITER toroidal field coil, which was completed in Italy in 2016.

schedule of ITER."[13] This, of course, only compounds ITER's cost and schedule problems. On the flip side, there have been painful cuts to the US domestic program. In 2016, the Alcator C-Mod tokamak at the Massachusetts Institute of Technology was shut down leaving the US with just two large tokamaks. The US was left with a weaker experimental tokamak program than the village of Culham in England.[14]

Nevertheless, the $25 billion price tag of ITER, since it is spread over half the world's population and across 25 years, is fairly small by governmental standards. Again taking the US as an example, the average citizen contributes about 20¢ a year to the project.

---

[13]From the FY 2018 Congressional Budget Justification for Fusion Energy Sciences.
[14]The US has DIII-D and NSTX, while the Culham Centre for Fusion Energy has JET and MAST.

This is less than peanut subsidies or a single F-35 fighter jet.[15] In fact, America's highest paid CEO makes enough to completely cover America's annual contributions to ITER. All of these examples are to say that ITER isn't a particularly high stakes issue for US politicians. While it appears that there are a few representatives that strongly support ITER, for the most part, it's a pawn that is tossed about in the political turbulence of Washington. Since fusion is such a long-term endeavor that the public is fairly unaware of, it's usually a tool that politicians use to achieve other policy aims (rather than being a policy aim itself). When Reagan and Gorbachev needed a symbol of East-West cooperation and a positive face for nuclear technology, they started ITER. When the US wanted to poke France for not supporting an invasion of Iraq, they supported Japan's bid for the ITER site. Currently, the device that could demonstrate the scientific feasibility of fusion is just one of many many poker chips on the table for politicians. While we object to this state of affairs, it's actually encouraging because it means that a small increase in public support could dramatically brighten ITER's future.

7.4 Transition to DEMO

Assuming that ITER meets its physics and technology goals, the next step would likely be a demonstration power plant. Anticipation for this device is so great that it has already been given a name: DEMO.[16] The purpose of DEMO will be to leverage the results of ITER in order to demonstrate that fusion can be economically viable. It will need to achieve slightly better plasma performance (i.e. $Q > 20$ with steady-state operation), but the most significant challenges will be in its engineering. DEMO will actually put net power onto the electric grid and, as such, must be much more reliable. It will need to operate continuously for very long periods of time (e.g. months).

---

[15] America's annual ITER contribution will average to $90 million, while a single F-35 fighter jet is estimated at $160 million and the peanut program is projected to cost $200 million annually between 2017 and 2019.

[16] Confusingly, DEMO is not an acronym, but is still capitalized (like ITER).

Moreover, because it operates continuously for so long, it must breed enough tritium to be self-sufficient.

Even now, while ITER is still under construction, tokamak designers have been busy planning for DEMO. For the most part, each country is developing their own design with the plan to build it by themselves.[17] For example, the European Union has designed the EU DEMO, which is a large, conservative machine with a radius of 9 meters. On the other hand, the Koreans are being more aggressive (and risky) in designing the smaller K-DEMO. They are pushing the limits hard because their government has set a particularly ambitious timeline to get to a power plant.[18]

The biggest wildcard in fusion energy is China. In 2006, the Chinese completed EAST, the first tokamak in the world to have superconducting toroidal and poloidal field coils. In the past decade, they have made a concerted effort to train enormous numbers of fusion scientists and engineers. China is developing its fusion workforce as it plans to construct its own ITER-like reactor: the Chinese Fusion Engineering Test Reactor. CFETR is planned to include two phases. In the first, it will have similar capabilities to ITER, but then it will be upgraded in order to serve as a stepping stone to DEMO. In this second phase, it aims to achieve $Q > 10$, $Q_{\text{eng}} > 1$, and a gigawatt of fusion power. While construction plans have not yet been finalized, China plans to begin operating phase I of CFETR in the early 2030s.

So why are the Chinese unilaterally investing in CFETR when they are already part of ITER? The idea is that, if fusion does become commercially viable, technological know-how and industrial capacity will be the key to exploiting it. By building its own reactor with the majority of the components sourced domestically, China will have a much more developed fusion industry than any other country, giving it a competitive advantage. With ITER, each member country only gains experience with a limited portion of the overall hardware

---

[17]The US stands out for displaying little interest in developing its own DEMO.
[18]As South Korea currently imports over 80% of its energy, one can understand the rush.

(although every member has access to all of the intellectual property). By building CFETR domestically, China can develop technology exclusively for its own benefit.

7.5 Other Things to be Excited for

Before we move on, we must acknowledge that 2025 (and 2035) is far away. Fortunately, there are a few exciting things to tide fusion fans over until ITER fires up. Watch out for these projects, coming to a fusion lab near you:

(1) **W7-X**, the world's largest and best optimized stellarator, was first switched on in December 2015. Following extensive testing, the first discharges with a divertor took place in September 2017. While W7-X recently broke the world record for stellarator performance, we won't truly see what the machine can do until upgrades are made in 2018 and 2020 to permit steady-state, high-power operation.

(2) **MAST-U**, an upgraded version of the existing MAST tokamak, is scheduled to commence operations in early 2018. But don't let the word "upgrade" fool you, it's a dramatically different machine. Most notably, it has the ability to investigate novel divertor configurations. Many of these show theoretical promise, but have never been experimentally tested. If MAST-U is successful, it could function with a divertor plate shielded by neutral gas, a game-changing solution to the power exhaust problem.

(3) **JET**, which achieved $Q = 0.7$, plans to perform a third set of deuterium–tritium experiments in 2019. As discussed last chapter, scientists aren't expecting to reach $Q = 1$. Rather, the primary aim is to support ITER and accumulate experience operating machines with a more robust first wall and divertor design. However, in fusion surprises are never too surprising and $Q = 1$ isn't that far away…

(4) **JT-60SA**, a new version of the JT-60U tokamak in Japan, is due for completion in late 2020. With an increased plasma volume of 130 cubic meters, it will surpass JET as the world's largest

tokamak. The device will be able to run steady-state discharges and its external magnets will be entirely superconducting. Most impressively, it should be able to create pure deuterium plasmas with the triple product necessary for $Q > 1$ in D–T plasmas. Unfortunately, since it won't have tritium management facilities, it won't be able to create D–T plasmas to actually generate the energy. In other words, JT-60SA should be able to create deuterium plasmas that, if tritium had been used, would have achieved $Q > 1$. Cool.

Chapter 8

The Future: Designing a Tokamak Power Plant

Designing a demonstration tokamak power plant, the next step after ITER, is a balancing act. As currently envisioned, it will depend critically on disciplines ranging from material science to electromagnetism, from fluid mechanics to thermodynamics. The device will span the full temperature range of the Solar System, with superconducting coils near absolute zero and plasma temperatures exceeding that of the Sun. It will be composed of thousands of different components, each with parameters to tune and constraints to satisfy. Many of these parameters are intimately connected. Modifying a parameter here can affect components all over the device. Yet, above all, a magnetic fusion power plant must be as simple, robust, and reliable as possible. This may seem to be an impossible challenge. *However, we will show that the multitude of disciplines involved and their interconnectedness is the greatest reason for optimism.*

To design an economic fusion power plant requires solving problems that involve a variety of independent disciplines in physics and engineering. Yet, because these problems are connected, progress in one can be used to compensate for shortcomings in others. Hence, the progress required in most individual disciplines is relatively modest. In other words, while designing a fusion power plant requires solving many problems, it has many more potential solutions. The goal of this chapter is to understand why.

8.1 Power Plant Design from First Principles

Let us start from first principles. Power plant design, like plasma physics, has a deceptively simple governing formula[1]:

$$\text{maximize}(\$), \qquad (8.1)$$

where $ is the profit made by the electricity company that builds and operates the plant. Any fusion power plant must be economically successful in the business of electrical utilities. To some it may seem that once we *can* build a fusion power plant, we *should* build a fusion power plant. After decades of anticipation, fusion may seem to justify itself. However, the harsh reality is that fusion is just another way of boiling water. And why would you do so by burning twenty dollar bills if you could use ones?

Equation (8.1) sets the ultimate goal. We will know that fusion has arrived when it stops being the exciting thing to do and starts being the cheap thing. Of course there are caveats to this, several of which significantly impact the viability of fusion. As mentioned in Chapter 1, the current electricity market does not properly account for several important externalities, most notably climate change, atmospheric pollution, and nuclear proliferation. Additionally, the current scarcity of energy resources drives global conflicts (e.g. wars over oil), which have very steep economic and human costs. The failure to account for all these externalities puts fusion at an artificial disadvantage because the externalities of fusion energy appear small compared to other energy sources. Thus, corrective measures (e.g. a government tax on carbon) would be much appreciated. Sadly, Equation (8.1) still holds regardless of whether the rules of the market are fair or not. Electric utilities seek profits in the real world, not our idealized world.[2]

---

[1] In free, capitalist energy markets.
[2] In 1995, the European Commision's ExternE project estimated the externalities associated with different energy sources. They concluded that price of coal generation would increase by 50%, gas by 20%, nuclear fission by 5%, hydroelectric by 5%, and wind by 2%. However, by the report's own admission, the uncertainties

While conceptually simple, Equation (8.1) is horribly impractical. Even just determining the future profits of a fully constructed plant is impossible. Here, we must consider Equation (8.1) for designs that only exist on paper. By making drastic approximations, we can employ this formula to motivate metrics that are more useful for our design work. As a first step, we expect a fusion power plant that does not have a cost of electricity (COE) that is competitive with other energy sources will generate little profit. Even with direct governmental support, we require fusion power plants to have economics that are at least in the ballpark of other sources. Hence, we arrive at a new formula that is somewhat less accurate, but more useful:

$$\text{minimize} \, (\text{COE}). \qquad (8.2)$$

This says that we must minimize the cost of electricity (i.e. $ per kWh) of a fusion power plant.

For the next step, we return to our explanation of tokamaks from Chapters 4 and 5 and ask "Which parts of this process cost the most money?" What we find is that the economics of fusion power plants look to be fairly analogous to nuclear fission plants. Both energy sources are very energy-dense, but harnessing them requires high-tech and sophisticated equipment. Because of this, we expect that operating costs (such as buying the fuel) will tend to be cheap compared to building the plant in the first place. In fact, only 10% of the COE from existing nuclear fission power plants is fuel costs, while 75% is up-front capital costs associated with building the plant.[3] Likewise, fusion power plants are expected to

---

were very large. For example, the effect of global warming was included by calculating the cost of *avoiding* significant climate change. This would seem to be the cheaper path, but not the one we are taking. Additionally, the cost of nuclear proliferation was not included due to the classified nature of the relevant data. Understandably, fusion was not considered.

[3] In fact, the most important factor in determining the financial viability of a fission power plant is often the interest rate on the loans used to finance construction.

cost billions of dollars to construct, while the fuel needed for a year of operation only costs tens of thousands. This contrasts with coal and natural gas, where capital costs only constitute 60% and 20% of the COE, respectively. With the assumption that the cost of a fusion power plant is dominated by its construction, we can approximate Equation (8.2) as[4]

$$\text{maximize}\left(\frac{P_{\text{elec}}}{\text{COP}}\right). \tag{8.3}$$

Hence, for fusion power plants, we want to find designs that minimize the construction cost of the power plant (denoted by COP), but generate the most electricity (i.e. net electrical power, denoted by P_{elec}).

Before we get to the science behind tokamak power plant design, we have one last step that is critical to produce a simple and practical formula for our optimization. We will assume that the capital cost of a power plant is proportional to its size. So doubling the volume of the power plant will double its cost. While this is intuitive (bigger things are made of more stuff and stuff costs money), it is easy to find examples where it is completely wrong. For example, a liter of bottled water is larger than a diamond, but is much cheaper. However, when narrowly applied in comparing similar objects, it can work well. For example, a two liter bottle of water typically costs about twice as much a one liter bottle of water. Comparing construction costs of previous fusion experiments (as well as detailed cost estimates of proposed designs) indicates that this works reasonably well. Still, we must be mindful of its accuracy. When actually designing a device, one would perform a detailed economic assessment after completing an optimization using this less-than-rigorous cost estimate.

---

[4]It seems like we should also factor in the lifetime of the power plant here. However, this doesn't turn out to be important as long as the lifetime is longer than a few decades. Investors, impatient as they are, will require construction costs to be recouped within 10 or 20 years, regardless of how much longer the plant can continue operating for.

By estimating the cost of a power plant design using its volume, we can transform Equation (8.3) into

$$\text{maximize}\left(\frac{P_{\text{elec}}}{V}\right), \tag{8.4}$$

where we have represented the volume of the power plant with V. This tells us to maximize the power density (i.e. power per unit volume) of the plant. Hence, we conclude that the power plant designs that would be most attractive to electric utilities (and society) are small, but still produce lots of electricity. Not exactly surprising, right?

These arguments tell us that we should do everything we can to maximize the net electric power density of our design, rather than any other quantity. You can think of the power density as being the ultimate performance metric for power plant economics, just like the triple product was the ultimate performance metric for a D–T plasma. With this concrete, quantifiable goal in mind, we can now conclude our discussion of economics and delve into the physics complexities of fusion power plants.

As a warning, this chapter is a bit more technical than previous ones — designing a fusion power plant isn't easy! Moreover, it ties together a lot of concepts that you may have learned about for the first time earlier in this book. So feel free to flip back to previous chapters to jog your memory and don't be discouraged if you find the next fifteen pages to be a challenge. After all, the fusion community has been struggling with it since the 1950s.

8.2 Maximizing Net Electric Power

In order to maximize the net electric power of our plant, we know that we should increase the fusion power produced by the plasma. However, we must also remember that a future power plant will still need substantial external power. Even if a tokamak plasma is ignited, power will be required to drive the plasma current. This external power must be minimized as it detracts from the net electricity that can be sold to the consumer. The balance of this can be seen in Figure 8.1, which shows the electricity that is needed by the plant

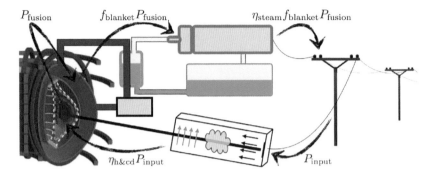

Figure 8.1: The flow of power through a tokamak power plant. Here, P_{input} is the external power required by the plant, $\eta_{\text{h\&cd}}$ is the efficiency of the heating and current drive systems, P_{fusion} is the fusion power produced by the plasma,[5] f_{blanket} is the power multiplication that occurs in the tritium breeding blanket, and η_{steam} is the steam cycle efficiency of the turbine.

and the electricity that is produced by the plant. Our ultimate power plant performance metric, the net electric power *density*, is simply the difference of the two divided by the volume[6]:

$$\frac{P_{\text{elec}}}{V} = \eta_{\text{steam}} f_{\text{blanket}} \frac{P_{\text{fusion}}}{V} - \frac{P_{\text{input}}}{V}. \tag{8.5}$$

To maximize the net electric power density, we obviously want the energy multiplication factor from tritium breeding reactions in the blanket, f_{blanket}, to be large. However, this is constrained by the energy produced from each lithium-6 reaction, so it is hard to imagine this value changing substantially. Additionally, we would like to increase the thermal efficiency of our steam cycle η_{steam} as much as possible, which can be done by increasing the operating temperature of the blanket/coolant. These two values, the blanket

---

[5] Everywhere in this figure, P_{fusion} should be replaced by $P_{\text{fusion}} + \eta_{\text{h\&cd}} P_{\text{input}}$ because the injected power doesn't magically disappear in the plasma. However, we will neglect it for simplicity because it will necessarily be small in a viable power plant. Additionally, while this figure is similar to Figure 5.11, we have written the power output in terms of the fusion power, rather than the plasma power multiplication factor Q. This revised notation allows for the possibility that the plasma is ignited (i.e. $Q = \infty$).

[6] Here, we will switch the definition of V to be the volume of the plasma, implicitly assuming that it is representative of the volume of the overall power plant.

power multiplication and the steam efficiency, are important, but largely independent from the rest of the power plant design.

The real optimization challenge comes in maximizing the fusion power density (i.e. P_{fusion}/V) and minimizing the external power density that must be drawn from the grid (i.e. P_{input}/V). These quantities depend strongly on the plasma parameters. We will delay discussion of the external power density until later and first focus on the fusion power density. Because of the form of the D–T fusion cross-section, the fusion power density has a simple dependence — P_{fusion}/V is directly proportional to the plasma pressure squared and nothing else.[7] While we haven't discussed the plasma pressure yet, it is the key to the economics of a power plant. *If you want to maximize the fusion power density, you have just one option: maximize the plasma pressure.*

8.3 Maximizing Plasma Pressure

Plasma pressure is simply the density of the plasma multiplied by its temperature (i.e. $p = nT$) and we have just said that it is key to achieving good power plant economics. This makes sense because, to maximize the number of fusion reactions in a given volume, you want to pack as many particles that are as hot as possible in it. Unfortunately, in existing experiments, as the plasma pressure is increased, eventually an MHD instability is triggered and the plasma disrupts. This constraint is formalized mathematically by what is called the *Troyon limit*. To understand the physics behind it, consider a tokamak to be a cage used to confine our fuel. However, instead of iron bars, it is made of magnetic fields. As we try to confine more particles (i.e. higher density) that each push harder on the cage (i.e. higher temperature), our cage is bound to fail. The Troyon limit predicts the maximum pressure our field lines can withstand.

---

[7]For D–T fuel with a temperature between 5 and 20 keV, we can approximate the value of $\langle \sigma v \rangle \approx c_{\sigma v} T^2$, where $c_{\sigma v} = 10^{-24}$ m^3/keV^2/sec is a constant. This is the same assumption we needed to derive the triple product in Chapter 4. The result is that $P_{\text{fusion}}/V = c_{\sigma v} E_{\text{fusion}} p^2 / 16$, where $E_{\text{fusion}} = 17.6$ MeV and p is the plasma pressure.

Unsurprisingly, the maximum achievable pressure increases with the strength of the magnetic field, but it also turns out to increase with the plasma current.

TECH BOX: The Troyon limit

The precise form of the Troyon limit is given by

$$p \lesssim 0.01 \frac{IB}{r}, \tag{8.6}$$

where $p = nT$ is the plasma pressure, n is the number density (i.e. the number of electrons per unit volume), T is the temperature, I is the electric current in the plasma, B is the magnetic field in the center of the plasma, and r is the radius of the plasma cross-section.

The Troyon limit is incredibly important because of the direct relationship between the plasma pressure and the fusion power density. Moreover, the triple product $nT\tau_E$, which we must increase to approach ignition, is simply the plasma pressure times the energy confinement time. Hence, it is clear that a power plant should push up against the Troyon limit to get the plasma pressure as high as possible. Unfortunately, a consequence of operating near the Troyon limit is an increased risk of disruptions. When our magnetic cage buckles from extreme plasma pressure, the plasma shoots off and touches the first wall. The heat and current from the plasma is immediately transferred into the solid material. While this process abruptly stops all fusion reactions, the energy and current stored in the plasma can still melt parts of the first wall and/or break off components. For this reason, we are confronted with a trade-off. Increasing the plasma pressure allows a given device to produce more fusion power, but it makes for less reliable operation and could cause significant damage. Thus, reactor design always involves a game of chicken against the Troyon limit — you push the limit as far as you dare.

The mathematical form of the Troyon limit tells us that increasing the plasma current will permit higher plasma pressure without exciting MHD instability. This motivates maximizing the plasma current, but, as we'll see in the next section, this turns out to be limited by an MHD instability as well. While operating within these limits may still allow for an economically viable power plant, some out-of-the-box thinking could reveal some attractive alternatives. It turns out that the Troyon limit can, in some circumstances, be violated. One promising method uses plasma flow, which is when the whole plasma has a net velocity, spinning around in the toroidal direction (see the following Tech Box). Experiments indicate that, with a fast enough flow velocity, the plasma pressure can be increased to roughly 50% above the Troyon limit without incurring MHD instabilities! While this appears to work, the hard part is finding a way to drive the flow (though some techniques are currently being researched).

TECH BOX: Violating the Troyon limit using flow

When the plasma starts to disrupt, it kinks (losing toroidal symmetry) and moves towards the first wall. As the plasma moves, it must move the magnetic field lines along with it. However, the first wall is metal, so the change in the magnetic field induces an electric current in the wall. This induced wall current then creates a magnetic field that reinforces the failing magnetic cage. Normally, this mechanism just slows the disruption down because, when the plasma moves slower, the induced current and hence the reinforcing magnetic field become weaker. However, if the whole plasma is flowing toroidally, then it can't go too slowly. The plasma has to race to push the magnetic field through the wall! Otherwise the plasma moves and everything has to start over because the plasma finds itself facing a different section of wall. In reality, this mechanism is even more complex than described here. Yet, it has a simple interpretation: if the plasma is flowing, then the solid first wall can help to support the magnetic cage, enabling it to hold back more plasma pressure!

8.4 Maximizing Plasma Current

We would like to maximize the plasma current because the Troyon limit tells us that doing so will enable higher plasma pressures. This would increase the fusion power density and, therefore, improve the economics of a power plant. However, when discussing MHD stability in Chapter 4, we learned that if the plasma carries too much current, the toroidal magnetic field won't be able to hold it in place and the whole plasma will kink. The onset of this instability is governed by what is called the *kink limit*. As with the Troyon limit, approaching the kink limit is risky because it can lead to a disruption, destroying the plasma and possibly damaging the surrounding components. Thus, in order to maximize the plasma current that can be stably achieved, we must increase the toroidal magnetic field generated by our external magnets.

TECH BOX: The kink limit

The precise form of the kink limit is a direct constraint on the safety factor $q \gtrsim 2.2$, which is equivalent to

$$I \lesssim \frac{1}{2.2} \frac{2\pi}{\mu_0} \frac{r^2}{R} B \qquad (8.7)$$

for a tokamak with a circular cross-section.[8] Here I is the electric current in the plasma, r is the radius of the plasma cross-section, R is the radius of the torus when looking down from above, B is the strength of the magnetic field, $\pi = 3.1415\ldots$ is the mathematical constant, and $\mu_0 = 4\pi \times 10^{-7}$ Henries per meter is a constant of nature known as the vacuum permeability. Substituting the kink-limited plasma current

(Continued)

---

[8] To derive this, start from the formula for the safety factor in a torus with a circular cross-section: $q = rB/(RB_p)$. Then, calculate the strength of the poloidal magnetic field B_p from the plasma current using Ampere's law. The presence of plasma shaping changes this derivation, resulting in a less restrictive kink limit.

(Continued)

(i.e. Equation (8.7)) into the Troyon limit (i.e. Equation (8.6)) yields

$$p \lesssim \frac{0.01}{2.2}\frac{2\pi}{\mu_0}\frac{r}{R}B^2, \qquad (8.8)$$

demonstrating the importance of a strong toroidal magnetic field.

If we assume that a power plant will be constrained by the kink limit, we can use it to calculate the maximum current. Plugging this result into the Troyon limit leads to a simple conclusion: *the maximum plasma pressure is entirely determined by how strong a toroidal magnetic field you can generate with your external coils.*[9] More specifically, the plasma pressure is proportional to the magnetic field squared. Hence, because the fusion power density is proportional to the pressure squared, this implies that the fusion power density is proportional to the magnetic field to the fourth power. While there are some caveats to this,[10] it clearly displays the crucial role of the magnetic field strength to the economic viability of a future power plant. Increasing the magnetic field by a factor of 2 would increase the fusion power produced by a given device by a factor of 16!

8.5 Maximizing Magnetic Field Strength

As is apparent from our discussion and the name "*magnetic* confinement fusion," high magnetic fields are very helpful for a power plant.

---

[9]Strictly speaking, it also depends on how large you can make r/R (i.e. how small you make the donut hole relative to the size of the donut). However, this is generally limited to about $r/R = 1/3$ by technological limits. You must have space for magnets and the tritium breeding blanket.

[10]For example, the toroidal field magnets are typically the most expensive component in the power plant. Thus, it is probably optimistic to assume that the construction costs would stay constant when increasing the magnetic field at constant plasma volume. Additionally, it isn't *necessarily* optimal to operate at the kink limit (as we will soon see) because it increases the external power needed to run the reactor. However, most power plant designs do.

We have just seen that a larger magnetic field relaxes both the Troyon and kink limits, which strongly improves the prospects for a fusion power plant. The limit on the maximum achievable magnetic field is set by one of two factors: the magnetic field strength allowed by the superconducting material or the mechanical stresses the toroidal field coils can tolerate.

In the early days of superconductivity, the materials could not function in high magnetic fields. If the field was too high they would lose their superconducting properties. Hence, magnet designers didn't have to worry much about the mechanical forces that made the coils want to blow themselves apart. The superconductors couldn't generate high enough fields for this to be a problem.[11] Since then, significantly better superconducting materials have been discovered (e.g. niobium–tin), which permit large enough fields to make mechanical stresses a real concern. Hence, modern tokamaks have "D"-shaped coils with a lot of structural support to best handle the stresses.

In the future, the balance could tip even further. The advent of high-temperature superconductors has the potential to largely eliminate the consideration of the field allowed by the material. Instead, we would be free to increase the magnetic field until the mechanical forces became completely insurmountable. This, of course, would be great for fusion as it would enable smaller power plants with better economics. The potential benefits have been illustrated in a recent tokamak power plant design: the ARC reactor [31]. ARC suggests that high-temperature superconductors *could* enable ITER-like performance in a JET-sized device (i.e. a burning plasma in a device with one-sixth of ITER's volume).

The important question that remains is "How far can we push this?" What are the maximum mechanical stresses that can be tolerated by large high-temperature superconducting magnets? This is a messy engineering question that depends on optimizing coil design as well as the detailed material properties of a given high-temperature

---

[11]The mechanical stresses in a toroidal field coil are proportional to rB^2.

superconductor. The best superconductors tend to have a complex response to the strength and orientation of the magnetic field they are immersed in. As a consequence, their performance can depend sensitively on seemingly inconsequential details like how the material is manufactured and how the wires are wound into coils. Yet, these details are at the heart of what may be the most important question for the future of fusion energy.

8.6 Minimizing External Power

Thus far, we have only concerned ourselves with the power *production* half of our net electricity equation (i.e. Equation (8.5)). However, the other half, the external power required to run the plant, has the potential to be a deal-breaker. We must ensure that it is small compared to the power production. There are two tasks that can dominate the power needed by our tokamak: plasma heating and current drive. Plasma heating is necessary in order to replace the energy that turbulence kicks out of the plasma. If the plasma is ignited, we don't need to worry about this because the lost energy is entirely replaced by fusion power generated in the plasma. If it isn't ignited, we must provide the difference ourselves using heating systems. In addition to this, we must drive a steady-state plasma current, which generally requires a lot of power. Plasma heating and current drive are closely related and are often accomplished together using neutral beam injectors and/or electromagnetic waves. Regardless, the power needed for both must be minimized.

8.7 Minimizing Heating Power

The goal of ITER is to create a burning plasma — a plasma that is largely heated by its own fusion energy. Specifically, ITER plans to achieve $Q = 10$, meaning the fusion power it produces will be ten times greater than the external heating power deposited into the plasma. Since four-fifths of the fusion power is carried by neutrons and immediately escapes confinement, ITER's self-heating will be twice as large as its external heating. In a power plant, the situation will necessarily be even more extreme. After all, ITER is designed

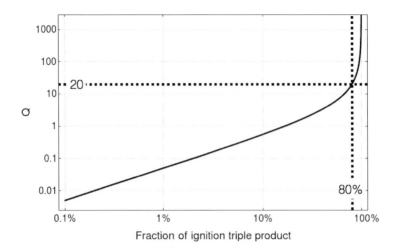

Figure 8.2: The value of the triple product needed to attain a certain plasma power multiplication factor. We see that, in order to achieve a large enough power multiplication factor for power plant, we need to be very close to the triple product that corresponds to ignition.

to be a science experiment, not to produce power. To produce a reasonable amount of net electricity, it appears that the plasma in a power plant will need to be at least $Q = 20$. Surprisingly, looking at Figure 8.2 (and also Figure 4.25 on page 136), we see $Q = 20$ is very close to ignition. In fact, at $Q = 20$, the triple product is 80% of what is needed for ignition at $Q = \infty$. This demonstrates that, once you get a high enough Q for any sort of net electricity production, the plasma is basically at ignition. Therefore, in order to minimize the heating power, we'll say that our power plant must ignite. This is just a bit overly stringent and it simplifies our analysis considerably.

In Chapter 4, we derived the ignition condition and found it was given by the triple product of

$$nT\tau_E \geq 5 \times 10^{21} \text{ keV-sec/m}^3. \tag{8.9}$$

In order to attain this value, we would like to increase the product of density and temperature as much as possible. However, this quantity is the plasma pressure $p = nT$ and we have already maximized it. Hence, we can think of the ignition condition as dictating the energy confinement time that we must achieve in our device. While ideally we would now turn to the physics of turbulent energy transport in order to explain how it can be obtained, we don't have a sufficiently deep understanding to do this. Instead, we must turn to the empirical scaling laws built from past experimental results. We discussed these in Chapter 4, but here we will write one out explicitly:

$$\tau_E = 0.20 \times I^{0.93} B^{0.15} P_{\text{fusion}}^{-0.69} n^{0.41} R^{1.39} r^{0.58}. \tag{8.10}$$

Looking at this long and complex expression, we don't see many unconstrained parameters that we can play with.[12] The plasma current I is already set by the kink limit, the magnetic field B has technological limits, and the total fusion power P_{fusion} is determined by the plasma pressure. In fact, we only have two free parameters: the plasma density n and the reactor size (set by the two radii of the donut R and r).

Since the pressure is the product of density and temperature and is fixed by the Troyon limit, increasing the density implicitly decreases the temperature. So the empirical scaling law is telling us that, for the best confinement, we should maximize the density because it allows us to minimize the temperature at constant pressure. In the next section, we will see that, like the pressure and the current, the plasma density is constrained by a disruptive limit: the *Greenwald limit*. For this reason, unless we can find clever ways to improve the empirical scaling or the aforementioned design limits, our only option to achieve the needed confinement is to increase the size of the device through R and r.

This is interesting because, in all our discussions of the fusion power density, the plasma's size wasn't important. It turned out that the effect of the size canceled, such that a small device could achieve the same power density as a big device.[13] Here though, after

---

[12] While it is neglected here for simplicity, plasma shaping also enters into the confinement time scaling. It should be increased as much as possible, but it too is constrained by the onset of MHD instability.

[13] There are some practical limitations to this statement that we've ignored in our analysis, which would likely ruin the power densities of very small devices (i.e.

substituting the Troyon, kink, and Greenwald limits into the empirical scaling law, we see that a larger device is the key to attaining the confinement time needed for ignition.[14] The intuition for this is simply that bigger devices have more volume per surface area, which means it is harder for energy to find its way out. Thus, since all other parameters are constrained, achieving the confinement needed for ignition determines the size of our device. As evident by ITER, the present-day state of fusion suggests that a power plant must have a plasma cross-sectional radius of at least 2 meters. In other words, big.

As a note of caution, we must keep in mind that turbulent transport in tokamaks is still fairly poorly understood. It would not be particularly surprising if the empirical scaling of Equation (8.10) were to give a somewhat inaccurate prediction (for better or for worse). Even amongst the experimental data used to generate the scaling formula, the confinement time can easily be off by more than 15%. Extrapolating to a power plant is even less certain because it will have parameters far outside of the range of what has been tried before. Experiments on ITER would go a long way in reducing this uncertainty. Additionally, physics-based models such as gyrokinetics can help. Unfortunately, while gyrokinetic simulations can model an individual design, they are still too expensive to be useful in large-scale conceptual design optimizations. Generally, crude and simple empirical scalings are used to find the operating point, which can then be assessed with more accurate, but expensive, tools like gyrokinetics. A second caveat is that new and innovative strategies for combating turbulence may be developed that improve matters. For example, before the H-mode was discovered, there was an L-mode scaling that was analogous to Equation (8.10). However, it had different dependences and predicted confinement times that were roughly a factor of two lower.

$R \lesssim 0.1$ meters). For example, it becomes impossible to prevent heavy impurity atoms from the first wall from getting into the plasma and radiating away energy.
[14] After plugging in all the formulas for the MHD limits, we find that the triple product is roughly proportional to the square root of the plasma's linear size as well as being proportional to $B^{0.75}$.

8.8 Maximizing Plasma Density

In order to maximize the energy confinement time, we would like to increase the density of particles in our plasma. Unfortunately, like the plasma pressure and current, there is a hard upper bound on the plasma density: the Greenwald limit. It is the least understood of all the design limits presented in this chapter. While the Troyon and kink limits are results of MHD calculations, the Greenwald limit was primarily motivated by experimental results. It had long been observed that, if you fuel a tokamak too aggressively, it will disrupt. What Greenwald did (building on the work of others) was to find a physically motivated rule of thumb that roughly matched the experimental results. In other words, by running many experiments, it was possible to determine how much density was too much and how this value depended on other plasma parameters. While the physics behind the density limit are still poorly understood, it is generally characterized by an increase in the amount of electromagnetic radiation produced by the plasma and/or a precipitous decline in particle confinement. The reasons for this behavior, though, remain largely a mystery.

TECH BOX: The Greenwald limit

The Greenwald density limit is given by

$$n \lesssim 10^{20} \mathrm{MA}^{-1}\mathrm{m}^{-1} \times \frac{I}{\pi r^2}. \tag{8.11}$$

Note that, unlike the Troyon and kink limits, the constant in the Greenwald limit has units, indicating its empirical nature.

8.9 Minimizing Current Drive Power

Thus far, we've developed a pretty tidy picture of power plant design. We simply minimize this parameter and maximize this other one ... and voilà, a power plant. Unfortunately, it isn't always that clear cut. If our device is large enough to come close to ignition, it won't need much heating, but driving current may still require substantial power. In fact, if we're not careful, driving the plasma current could easily consume all the electricity produced by the power plant. For this reason, it's very important to minimize the externally driven current and maximize the efficiency of current drive techniques (e.g. how many units of current are driven per unit of external power).

Immediately, we see that this first motivation, minimizing the amount of current, directly contradicts our previous optimization. Earlier, motivated by higher plasma pressures and better confinement, we increased the current as close to the kink limit as we dared. Now, we are saying that we want to decrease the current. This is an important trade-off in power plant design. Driving more current requires more external power, but enables smaller devices with higher fusion power densities. An attractive option to minimize the *externally driven* current without hurting plasma performance is to maximize the current that is self-generated by the plasma — the bootstrap current. Increasing the bootstrap current allows a device to have the same amount of total current, while needing less external power to drive it. Some power plant designs rely on the bootstrap effect to multiply the externally driven current by a factor of 10! This, in turn, reduces the power needed to drive current by a factor of 10. While a modest bootstrap multiplication (a factor of two or three) is definitely beneficial, going too far is difficult to accomplish and can cause other problems. To see why, consider an ignited tokamak where only 10% of the current is externally driven — the plasma is almost entirely self-sustained, meaning that operators have very little control over it. Instead of being able to optimize the plasma to our wishes, it will do whatever it wants. This lack of operator control seems likely to reduce plasma confinement and increase the risk of disruptions. Hence, the optimal amount of bootstrap multiplication is very much an open question.

The second motivation, maximizing the efficiency of current drive, further complicates matters. There are several different ways to drive current (e.g. neutral beam injection and a number of distinct techniques using electromagnetic waves), each of which have different dependences on plasma parameters. However, generally they all become less efficient at higher densities because the electrons that carry the current collide more and slow down quicker. This conflicts with our desire to operate at the Greenwald limit in order to maximize the energy confinement time. Hence, we are forced to prioritize either confinement or current drive efficiency.

In summary, our desire to minimize the external power needed to drive the plasma current conflicts with several of our previous optimizations. Nevertheless, most present-day power plant designs tolerate the increase in external power consumption in order to operate at the kink and Greenwald limits. This is for good reason. Detailed analysis indicates that doing so does indeed lead to the lowest cost of electricity. However, this is a reflection of the present state of fusion research and need not always be the case. The future balance really depends on what progress is made in:

- the efficiency of current drive systems and the bootstrap current,
- the ability to withstand/mitigate disruptions, and
- the energy confinement time.

If scientists discover a way to quench turbulence, making ignition easily attainable, then why bother driving the maximum amount of plasma current? It would unnecessarily consume extra power and increase the risk of disruptions.[15] On the flip side, if a super-efficient new current drive method is developed, it would further reinforce the desire to drive as much current as possible within the MHD stability limits.

8.10 Maximizing Material Survivability

Finally, there is one last consideration that we have yet to mention: the survivability of the fusion technology around the plasma. We have seen that confinement is the main consideration that determines the size of the device. However, our discussion seemed to suggest that the

---

[15]Though it would still help you achieve higher fusion power densities, which might make it worth it.

reactor size could be increased without limit. This is not true. If you make the device larger while keeping the plasma pressure constant, the total power produced by the plasma increases in proportion to the plasma volume. Bigger plasmas make more power. However, whenever this power comes out of the plasma, it can only come out through a surface. In fact, this is the exact same reason why bigger plasmas have the better confinement that we need. They have a larger volume to surface area ratio. However, the downside is that this increases the power per unit area that the solid components surrounding the plasma must tolerate. There is one specific case that illustrates this well: divertor heating.

In Chapter 5, we learned about the divertor, a specialized part of the first wall that handles the power that continually leaks out of the plasma. Because particles travel quickly within magnetic surfaces and move slowly across them, the escaped power is deposited in a very thin toroidal line on the divertor plate just outside of the magnetic surface with the X-point. Now let's say that we have a tokamak plasma with a certain volume that is exhausting its power into a thin toroidal line with a certain area. If we were to make both radii of the donut twice as large, the volume would increase by a factor of 8, as would the power leaving the plasma. Now, the area over which this power is exhausted does increase, but not by as much as you might think. The length of the thin toroidal line increases by a factor of two. Additionally, it would seem like the thickness of this line should double as well, but this actually turns out to be extraordinarily unclear. Predicting how the width of this heat deposition area changes with tokamak parameters is one of the most difficult problems in fusion. Thus, it is a big source of uncertainty in reactor design. Regardless, let's be generous in our extrapolation and imagine that the thin width of the line does double. This still means we are exhausting 8 times as much power into an area only 4 times larger, making it twice as difficult for the material to survive. If we were to triple the reactor radii, we would be exhausting 27 times as much power into an area only 9 times larger, a factor of 3 worse for material survivability. In other words, if we increase the reactor size enough, we are guaranteed to eventually melt our divertor plate. When the details of the divertor power handling are analyzed (i.e. combining the most accurate extrapolations with the current capability of divertor technology), we find that ITER looks to be fairly close to the limit of what is achievable. That being said, until we test ITER, the uncertainty will remain quite large.

Reducing the uncertainty in material survivability is especially important for a power plant. Unlike experiments, a power plant must provide a dependable and reliable supply of electricity. If a 1 GW power plant unexpectedly switches off, it can destabilize the electrical grid and cause power outages. Moreover, for every day that the plant is shutdown for repairs, the utility misses out on over a million dollars worth of electricity sales. Thus, the monetary value of reliability should not be underestimated. Conservative designs that operate far from material (and MHD) limits could very well yield more profits than those that push performance, especially given that the robotic maintenance needed in a fusion power plant appears time consuming and expensive.

8.11 Striking the Right Balance

You may have found the deluge of different quantities and motivations in this chapter hard to keep track of, so we have summarized the main points in Figure 8.3. What we see is that we can distinguish four motivations that together determine our reactor's plasma parameters:

- fusion power production,
- heating power consumption (i.e. confinement),
- current drive power consumption, and
- material survivability of the fusion technology.

Ideally, we would have excellent confinement, perfectly efficient current drive, and invincible materials. Under these conditions, the design optimization would be simple. You would do everything you can to achieve the highest possible magnetic field in order to relax the Troyon and kink limits and maximize the fusion power density of your plant. That would enable the smallest power plant to produce the most electricity. Unfortunately, we aren't that good at fusion yet, so a bunch of trade-offs arise.

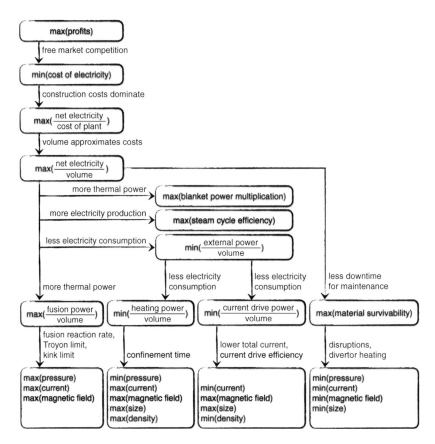

Figure 8.3: The motivations and optimizations involved in the design of a tokamak power plant. For example, the first two boxes indicate that the desire to maximize profits motivates minimizing the cost of electricity because of free market competition.

Reactor size is one example. Because turbulence degrades our plasma confinement, it sets the *minimum* size of the plasma in order to get near to ignition. Moreover, larger plasmas are preferable as they tend to require less external power to drive the plasma current.[16] On the other hand, the plasma also has a *maximum* size, above which

---

[16]This is primarily because the kink limit forces the total current *per unit of plasma volume* to decrease. In other words, it requires less external power per unit volume to reach the kink limit.

its power exhaust would destroy even the most durable divertor.[17] While there is still a lot of uncertainty in these two constraints on size, it's clear that they will become mutually exclusive if we push the fusion power density too high.

There are many other examples of trade-offs. Plasma current helps to achieve high fusion power densities and good confinement, but driving it requires a lot of external power. Decreasing the density improves the current drive efficiency, but doing so would lower the energy confinement time. Even increasing the magnetic field, which is a key to improving the economics of a tokamak, has a few drawbacks. It causes the electromagnetic forces in a disruption to get stronger and, by enabling a higher fusion power density, means that the divertor must handle more power.

In this chapter, we have tried to simplify power plant design as much as possible. However, we see that it is a messy engineering endeavor. We seek an optimum design amongst several competing considerations (e.g. fusion power density, confinement, material survivability) by varying a large number of independent design parameters (e.g. pressure, density, magnetic field). To account for all these practicalities, computer programs called *systems codes* are used. They have sophisticated estimates for the cost of electricity that they attempt to minimize by varying the design parameters, all the while ensuring that a large set of design limits are respected.

The advantage of all this complexity is that there are several avenues by which we may make progress. As an example, let's imagine that super-duper H-mode is discovered and improves confinement far past what is needed, while research on current drive hits a brick wall and makes no progress. In our power plant design, we could reduce the plasma density (and no longer operate at the Greenwald limit) in order to improve our current drive efficiency at the expense of confinement. As another example, imagine that scientists learn

---

[17]Additionally, large power plants can become prohibitively difficult to finance. Nuclear *fission* plants cost about $10 billion, which is near the limit of what the largest electric utilities can handle. Even so, they generally secure special government loans in order to make the financial risk of such a big project tolerable.

how to mitigate disruptions and build more robust divertors. Then, we could increase the size of our device and compensate for a lack of progress in confinement and/or current drive. Imagine that high-temperature superconductors enable stronger magnetic fields and innovative gas-shielded divertors become practical. Then, the fusion power density could be increased, the power needed for current drive could be reduced, and ignition could be achieved in smaller devices. These examples illustrate the primary point of this chapter: *by modifying the balance of design parameters, an advance with respect to one consideration can be used to compensate for lack of progress in other considerations.*

Finally, when designing a fusion power plant you are forced to confront one last trade-off — a meta-trade-off you could say. How much risk should you take? How bold should you be? In fusion, uncertainty abounds. The performance of these devices is sensitive to many physical mechanisms that are not well understood. Ambitious and creative designs will always hold the potential for better performance, but also a greater possibility of failure. In 1973, the JET tokamak was designed with a "D"-shaped plasma cross-section. This had never been tried before, but turned out to dramatically improve performance. On the other hand, in 1985, the JT-60 tokamak began operation with its magnetic X-point positioned on the outboard side of the plasma (instead of above or below it). Just two years of experimentation revealed that this degraded plasma performance and was unworkable. JT-60 needed to be retrofitted with new coils in order to move the divertor to its conventional location.

With many of these design decisions, the consequences will never really be known until they are tried. As the field of fusion matures, devices like ITER will gradually unravel many of these mysteries. At some point, a path to economic fusion energy will be found (whether it be a tokamak or something else). Yet, plasma physics and fusion are such rich and deep fields that, even after power plants are commonplace, the potential for improvement will remain.

PART 4
SPECIAL TOPICS

Chapter 9

Alternative Approaches to Fusion Energy

Thus far, this book has focused almost entirely on tokamaks. While it currently looks likely that first-generation commercial fusion power plants will be tokamaks, it is important to research a variety of approaches. We are still in the early days of fusion, and it seems much too soon to fully commit to any single device. For all we know, the ultimate fusion power plant could well be a stellarator ... or perhaps something more exotic.

9.1 Stellarators

Stellarators predate tokamaks. They were the brainchild of Lyman Spitzer, a visionary American physicist who founded fusion energy research in the United States.[1] However, after the incredible experimental performance of the Russian T-3 tokamak in 1968, most countries put their stellarator program on the back burner. Decades later, as it became clear that tokamaks still required a lot of work, stellarators began to experience something of a renaissance. Now they are again considered a legitimate competitor to tokamaks and, in many ways, have a more promising future.

First, let's briefly review our discussions from Chapter 4 about what distinguishes stellarators from tokamaks. Both are toroidal magnetic confinement devices, meaning they confine the plasma

---
[1]Spitzer was also a key figure in the conception and development of the Hubble Space Telescope.

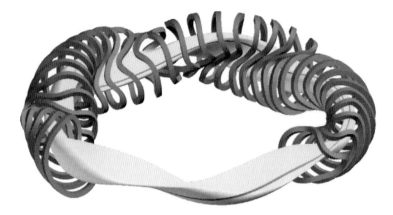

Figure 9.1: The coils and plasma in the Wendelstein 7-X stellarator (some coils have been removed for viewing clarity).

using a magnetic geometry with a hole in the middle. However, the two devices use different techniques to create the poloidal magnetic field (which is needed to neutralize the particle drifts and keep the toroidal plasma from destroying itself). The tokamak uses an electric current running through the plasma, while the stellarator uses external coils bent into funky shapes. Importantly, you can show mathematically that, for the stellarator to avoid having a plasma current, it must forgo toroidal symmetry. This means that a stellarator looks very different as you move around it toroidally (as shown in Figure 9.1). The magnetic field, the plasma density, the plasma temperature, and the poloidal cross-sectional shape of the magnetic surfaces can all vary. In contrast, a tokamak is a lovely, simple donut shape that looks the same as you move in the toroidal direction.

These two facts, that stellarators don't need a plasma current, but lack toroidal symmetry, are all you really need to know to understand how they compare to tokamaks. Not needing a plasma current has a bunch of positive consequences, while lacking toroidal symmetry has a bunch of negative consequences. We'll start with the bad news.

The most direct effect of breaking toroidal symmetry is on our understanding. Humans, with our puny little minds, have much more trouble thinking about the three-dimensional stellarator, compared to the two-dimensional tokamak. Stellarators are just intrinsically more difficult to analyze because the magnetic field is so much more complex.

A second disadvantage of breaking toroidal symmetry is that the particles that are trapped by the magnetic mirror force are no longer guaranteed to execute *closed* banana orbits (see page 118 for a reminder). If the magnetic field is not carefully designed, as the particles reflect back and forth between regions of high magnetic field strength, they will drift out of the device. This means that, in principle, neoclassical transport can exceed turbulent transport, lowering the energy confinement time.[2] Fortunately, neoclassical transport is fairly well understood (at least compared to turbulent transport). Hence, it has recently become possible to use supercomputers to optimize the magnetic fields to reduce neoclassical transport. One of the first such "optimized" devices is W7-X (see Figure 9.1), a JET-scale stellarator that began initial operation in Germany in late 2015.

A third difficulty concerns magnetic surfaces, which are vital to achieving good plasma confinement. A tokamak, due to its toroidal symmetry, automatically has nice, nested magnetic surfaces. However, the funky coils of a stellarator can be designed to create all sorts of magnetic fields, most of which are just a tangled spaghetti of magnetic field lines that don't form surfaces at all. Instead, the coil shapes must be designed such that they lead to well-formed magnetic surfaces with minimal magnetic islands and stochastic regions. Figure 9.2 indicates how hard this is. We see that even the W7-X stellarator has some magnetic islands (although they are quite narrow radially). These islands give particles a short-cut in their escape because, by traveling along the magnetic field around the island, they can move closer to the edge of the plasma.

Thus far, we have only considered the consequences of breaking toroidal symmetry for the plasma, but fusion technology is made more difficult too. In a tokamak, all the toroidal field coils are

---

[2] This loss mechanism is expected to be particularly severe for the high energy helium nuclei produced by fusion. It is important that future stellarator power plants are able to confine them, otherwise the plasma won't effectively self-heat.

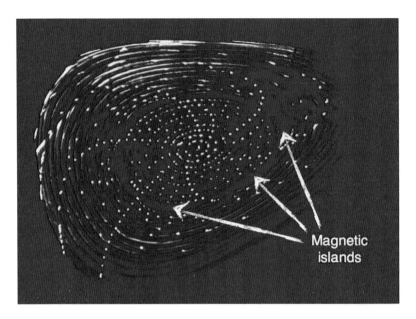

Figure 9.2: An experimental image of the poloidal cross-section of W7-X. For the most part we see nice, nested magnetic surfaces, except for the thin chain of six magnetic islands halfway out (the long dark blobs).

identical. They are flat and "D"-shaped to minimize mechanical stresses. Moreover, the vacuum vessel and many internal components are toroidally symmetric. These properties simplify the design of a device, its construction, and its maintenance. In contrast, the different magnetic field coils of a stellarator often have wildly varying shapes. They are three-dimensional and have sharp bends in them. This means that the coils in a stellarator aren't as robust mechanically, which reduces the maximum achievable magnetic field strength.

Despite all these difficulties, stellarators still hold the potential to make a very attractive power plant, primarily because they don't require any plasma current. The magnetic field can be entirely created by external coils. This means that stellarators are intrinsically steady state. No need to worry about inductive versus non-inductive current drive or maximizing the bootstrap mechanism. Consequently, it's easy to imagine a stellarator achieving a very high power plant multiplication factor. Consider an ignited stellarator plasma confined by superconducting coils. No external power would be required to heat the plasma, no external power would be required to run the magnets, and no external power would be needed to drive current.[3] Perhaps even more importantly, stellarators don't suffer from disruptions and a plethora of other instabilities that are driven by the plasma current. Tokamaks tend to have *hard* stability limits. This means that when you violate them, it causes a disruption and destroys the plasma. Instead, stellarators tend to have *soft* stability limits. If you try to say, confine too much fuel, it will excite instabilities that kick the excess fuel out, but without destroying the entire plasma. This would dramatically improve the reliability of a power plant and ease the material survivability concerns of the first wall and the divertor.[4]

Finally, a tantalizing property of stellarators is the sheer number of possibilities for the shape of the magnetic field lines. The parameter space of stellarators is absolutely massive and, relative to the tokamak, there has been little exploration of it (both experimentally and theoretically). Doing so is a complex and expensive task, but one that could substantially improve stellarators. Previously, we've said that our limited understanding of fusion devices is a reason for optimism because it holds the potential for improved performance. This is especially true for stellarators.

So, theory aside, how well do today's stellarator experiments actually operate? To measure plasma performance, we'll return to our reliable figure of merit — the triple product. Unfortunately, there haven't been many big stellarator experiments, so the comparison to tokamaks isn't entirely fair. With that caveat in mind, Figure 9.3 shows that stellarators have somewhat lackluster triple products.

---

[3]Power would still be needed for plasma control systems and to keep the superconducting magnets cold, which could very well be the largest requirements for external power. This would be great because this power is tiny compared to what would be needed to drive the plasma current in an ignited tokamak.
[4]However, remote maintenance using robotics is much more difficult in a stellarator due to its complex geometry.

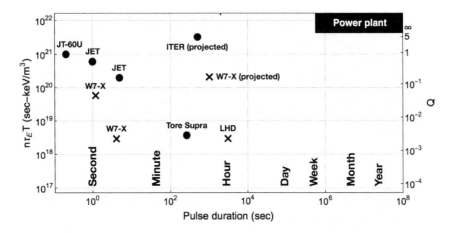

Figure 9.3: The triple product versus the duration of the plasma discharge for a number of experiments (as well as the projections for future performance). Circles indicate tokamaks, while the crosses indicate stellarators.

They're similar to what tokamaks were obtaining in the 1980s. On the other hand, we see that stellarators can easily outlast the longest pulse durations of tokamaks. Pulse duration, not to be confused with the energy confinement time τ_E, is the total duration of the plasma discharge. A commercial magnetic confinement fusion reactor will likely need to operate for weeks, months, or even years without a break. Because stellarator experiments are inherently steady state, it is much easier for them to run for long times.

In the near future, the stellarator concept will be subjected to some important tests. The performance of the W7-X experiment, relative to projections, could dramatically alter the future of stellarator research — for better or worse. As of late 2017, W7-X has been successfully constructed, which is an important and non-trivial step. Initial operations have begun and have achieved encouraging results (with triple products shown in Figure 9.3). However, before the machine can start high-performance discharges, a water-cooled divertor must be installed. This is scheduled to occur in two stages in 2018 and 2020. Soon afterwards, W7-X could do a great deal to demonstrate the viability of stellarators and inspire successive devices.

9.2 Inertial Confinement Fusion

For the most part, all of the confinement schemes we have discussed rely on using magnetic fields to control particle motion. This is referred to as MCF, or magnetic confinement fusion, in order to distinguish it from a second, fundamentally different approach to confinement that we have yet to discuss. This second approach is called *inertial confinement fusion* (or ICF) and, historically, it has comprised a significant fraction of fusion research and funding. Instead of using strong magnetic fields to confine the plasma in steady state, ICF uses powerful lasers to implode small capsules of fusion fuel.[5] By doing this, ICF attempts to achieve the triple product needed for D–T ignition, but with a very different set of parameters than MCF. Magnetic confinement fusion aims for a temperature of roughly 10 keV, a confinement time of several seconds, and a density of 10^{20} particles per cubic meter (about a million times *less* dense than air). ICF, on the other hand, aims for the same temperature, a much shorter confinement time of 1 nanosecond, and a much larger density of 10^{31} particles per cubic meter (about a million times *more* dense than air). This density is equivalent to two Boeing 747 airplanes crammed into the volume of a gym locker!

Figure 9.4 shows the different stages in a typical ICF implosion. You start with a tiny 1 millimeter spherical capsule comprised of a plastic shell filled with D–T fuel. The first step is to hit the capsule from all sides (and as uniformly as possible) with electromagnetic waves. As the waves hit, they heat the plastic into a plasma and blow it off the surface of the capsule. Due to Newton's third law — every action is balanced by an equal and opposition reaction — whatever remains implodes. In other words, the outer layers of the capsule get blown outwards, so the inner layers must get blown inwards. As the fuel implodes, the center ultimately becomes very hot (at least 10 keV) and very dense (around 1,000 times its original density). Once these conditions are achieved, so many fusion reactions will occur that the fuel at the center of the capsule will ignite. As long

---

[5]Alternatively, instead of lasers, particle beams have been proposed.

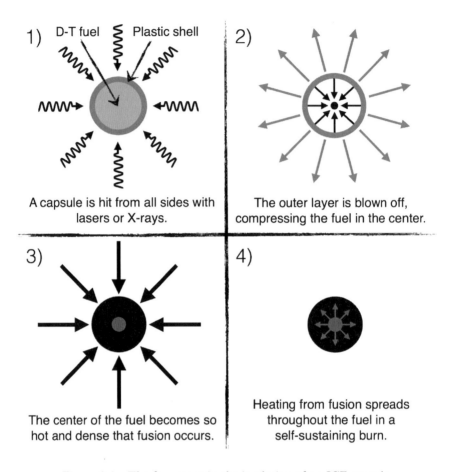

Figure 9.4: The four steps in the implosion of an ICF capsule.

as the fuel is sufficiently dense, the helium-4 nuclei produced by fusion will deposit their energy in the surrounding fuel, causing it to ignite as well. Of course, since there is no force continually providing confinement, after compression finishes, the plasma will rapidly expand and come apart. In order to generate electricity, this process will be repeated several times per second. The neutrons and heat from the mini-explosions can be used to breed tritium, boil water, and run a steam turbine.

To summarize, you can think of ICF as an attempt to adapt a thermonuclear hydrogen bomb for electricity generation. Two important changes are required: make the explosion small enough for a vacuum vessel to survive and ignite the fusion fuel using a laser instead of a nuclear fission bomb.[6] Don't fear though, ICF does not look at all practical to weaponize. The necessary laser facilities cost billions of dollars and fill buildings that span the area of several football fields. Conventional hydrogen bombs are already much smaller, cheaper, and easier to build.

The key difficulty in ICF is compressing the fuel by a factor of 1,000 to obtain the required density. To see the problem, imagine you are holding a small dough ball in your hands. Your task is to compress it uniformly, making sure that it always remains a sphere. This is very hard to do. Unless you can somehow squeeze exactly evenly, the dough ball will squish out between your fingers and make a massive mess. Physicists working on ICF face a similar problem! As the capsule is being compressed, the D–T plasma squishes out between the laser beams. This is known as the *Rayleigh–Taylor instability* (see Figure 9.5) and it is fairly common in nature (e.g. mushroom clouds from large explosions, astrophysical nebulae, and salt domes in geology).

In order to minimize the Rayleigh–Taylor instability and best compress the plasma, ICF researchers strive to create extremely smooth and spherical capsules as well as extremely uniform laser

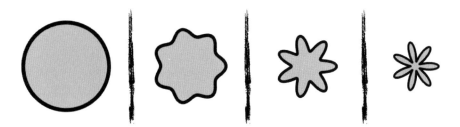

Figure 9.5: A four frame cartoon of the Raleigh–Taylor instability growing during an ICF capsule implosion. In this example, it looks like the laser is pushing slightly harder in seven places around the capsule. Though not shown, as time progresses, the instability becomes increasingly turbulent and chaotic.

---

[6]If you're not familiar the inner workings of a thermonuclear bomb, we'll review them in the next chapter.

systems. Any small imperfection in the capsule shape will rapidly grow during compression and prevent high densities. For this reason, manufacturing techniques have been developed to produce capsules that are smooth and spherical to one part in ten thousand! This is so precise that fabrication must take place in a clean room, otherwise dust particles landing on the capsules would ruin their smoothness. To illuminate the capsule as evenly as possible, scientists use a technique called *indirect drive* (see the right side of Figure 9.6). Instead of directly shining the lasers on the capsule (referred to as *direct drive*), they shine the lasers on a gold can called a *hohlraum* that surrounds the capsule. The laser causes the gold to emit X-rays, which enables more uniform illumination of the capsule.[7] Of course, the downside is that not all the laser energy incident on the hohlraum gets converted into X-rays. Hence, indirect drive provides a more even compression, but less total force driving the compression.

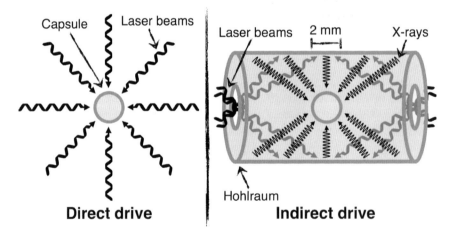

Figure 9.6: The two main methods of imploding an ICF capsule: direct drive (left) and indirect drive (right). In direct drive, the laser beams directly hit the spherical capsule. In indirect drive, the beams hit the inside of a gold can called a *hohlraum*, which then releases X-rays that hit the capsule.

---

[7]Additionally, X-rays have a shorter wavelength than the laser light, which, for subtle reasons, directly reduces the growth of the Rayleigh–Taylor instability.

Figure 9.7: (Left) The National Ignition Facility target chamber. 192 lasers converge on a target the size of a pea in order to create conditions similar to those within a nuclear weapon. (Right) The NIF laser bay. The peak laser *power* is 500 terawatts — about 300 times greater than the average power used by all of humanity.

So, how well is ICF doing? The world's largest and most energetic ICF facility is the National Ignition Facility (or NIF for short), located in Livermore, California. As shown in Figure 9.7, it is built on an awesome scale. It was completed in 2009 for a total construction cost of about $4 billion (adjusted to 2017 dollars), which lies in between the $0.5 billion JET experiment and the $25 billion projected price tag for ITER. As its name suggests, NIF has the goal of achieving fusion ignition, and it was designed to do so using indirect drive. Unfortunately, NIF's ignition campaign formally ended in 2012, without reaching its goal. The highest performing shots produced about 60 kilojoules of energy, just shy of the 100 kilojoules of external heating absorbed by the capsule. Why did NIF fail? The reasons are complex, but come down to plasma instabilities being more severe than expected. Instabilities like the Rayleigh–Taylor (and turbulence created by laser-plasma interactions) messed up compression and caused the plastic shell to partially mix with the D–T fuel, cooling it down and getting in the way of fusion.[8]

---

[8]Despite NIF's failure to ignite, it is still a fascinating physics experiment. For example, it can replicate some of the processes that occur in the cores of supernovae.

While NIF did not ignite, it still came relatively close. In fact, it achieved a higher triple product than any tokamak. Does this mean that we could just build a more powerful laser and have a power plant? A bigger laser could almost certainly achieve ignition, but would still be difficult to turn into a power plant. The reason for optimism is that, unlike with magnetic confinement, we already know what success looks like. ICF ignition was first achieved in 1952 in the context of a hydrogen bomb. Moreover, circa 1980, the US conducted a series of proof-of-principle experiments for the ICF concept. Dubbed Halite-Centurion, these experiments used the underground detonation of a nuclear weapon to generate X-rays, which were then used to compress a nearby ICF capsule. According to reports, ignition was successfully achieved on multiple occasions. However, the important question for ICF is "What is the minimum amount of X-ray energy that can still achieve ignition?" While many details are classified, one report claims that a 10 megajoule laser would enable an ICF system to simply mimic the Halite-Centurion results [32, 33]. Other reports claim higher values, some as much as 100 megajoules [32, 34]. In comparison, NIF has a capacity of just 1 megajoule. Nevertheless, the knowledge that ICF ignition is possible is compelling.

The reasons for pessimism are numerous. If we just built 100 copies of NIF and aimed them all at the same target, Halite-Centurion tells us it would almost certainly achieve ignition. However, it wouldn't make a power plant. While ignition has the same meaning in ICF and MCF, it has different consequences for energy production. A *very* careful reader may have noticed something a bit puzzling about the triple product numbers at the beginning of this section. While MCF aims to operate at ignition (or a bit shy), ICF aims to go beyond the ignition triple product by a factor of between 10 and 100. Why is ICF so ambitious? Because ignition isn't enough for net power production. In MCF, you have to put in some initial amount of energy to heat up the fuel and ignite, but then you can operate for months or years. Hence, the energy required to reach ignition is negligibly small compared to the total energy produced.

In ICF, however, ignition only lasts for about a nanosecond, after which you have to start over again and heat up a new fuel capsule. Thus, ignition isn't necessarily sufficient — you must ensure that the energy produced *during* ignition is significantly more than the energy needed to *reach* ignition. For this reason, to achieve the same plasma power multiplication factor Q, an ICF power plant would need a higher value of the triple product than an MCF power plant.

Perhaps the biggest issue in translating a device like NIF into a power plant involves fusion technology. While magnetic confinement systems have magic on their side (i.e. superconductivity) and relatively efficient heating systems, NIF has more losses. As an illustration of this, Figure 9.8 follows the *heating energy* from the electrical grid to the fuel in both NIF and JET. The difference is stark. In JET, about 25% of the electricity drawn from the grid for

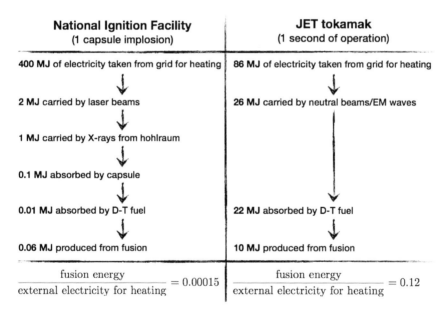

Figure 9.8: A rough comparison of the amount of energy required for the heating systems in the best ICF (left) and MCF (right) experiments to date [20, 35, 36].

heating is absorbed by the D–T fuel.[9] In NIF, only about 0.0025% gets to the fuel. Thus, for the same plasma Q, NIF is a factor of 10,000 further away from an energy producing power plant. Moreover, while NIF can only fire its lasers a few times *per day*, an ICF power plant would need to shoot capsules a few times *per second*. Achieving this amount of improvement requires some advances in laser technology. Finally, we have seen that, in order to maximize compression, NIF uses a gold hohlraum and capsules that are extraordinarily spherical. Unsurprisingly, they don't come cheap. Currently, NIF buys each hohlraum/capsule combo for about $2,500, while it is estimated that a power plant could afford to pay no more than $0.25 (they could buy in bulk though!).

To summarize, it currently looks like an ICF power plant would require a compression system that is 10 times more powerful than NIF, 10,000 times as efficient, fires 100,000 times more quickly, and implodes targets that are 10,000 times less expensive. While we should remember that NIF wasn't optimized to be similar to a power plant, the sheer magnitude of all these numbers is not encouraging. That being said, in Chapter 2, we learned that all of our sustainable energy options have substantial roadblocks. Just because ICF looks difficult doesn't mean we shouldn't try. For example, in the same lab as NIF is the Mercury laser, which is aimed at developing laser technology for an ICF power plant [37]. While much less energetic than NIF, it can fire at the required rate of 10 times per second and is a factor of 10 more energy efficient.

Still, if the road to a power plant looks so difficult, then why was the US willing to spend $4 billion to build NIF? Well, first off, before NIF obtained its results, ICF looked somewhat more promising. When NIF was designed, complex computer codes (validated by the Halite-Centurion tests) indicated that it would ignite. The

---

[9]Since JET has copper magnets, they consume a large amount of power that we are ignoring here. However, it seems fair to assume that JET could have been operated with superconducting coils because JT-60U is superconducting and has achieved even better plasma performance (without tritium). Alternatively, you could imagine installing JET's tritium handling facilities on JT-60U.

unexpected strength of plasma instabilities may impact future ICF funding. Second, fusion energy research is just one of the purposes of laser compression systems. A big part of NIF's mission is to study the conditions found within nuclear weapons, which is why it's not surprising that it was entirely paid for by the US National Nuclear Security Administration. NIF allows the US military to investigate the compression that occurs in the center of hydrogen bombs, but without violating nuclear test ban treaties by detonating a nuclear weapon. Moreover, NIF helps to maintain a large, active scientific workforce interested in weapons-related physics. The best evidence for this can be seen in France. Despite NIF's failure to achieve ignition, the French have pressed on with the construction of Laser Mégajoule, a slightly less powerful version of NIF. Its main objective is nuclear weapons stewardship, though it also performs ICF research. While it's easy to demonize the weapons relevance of ICF, it's actually a great asset that should be leveraged as much as possible. The funding advantages that ICF enjoys may well allow it to overcome the disadvantages it was dealt by Mother Nature.

9.3 Private Fusion Startups

In the 1980s, due to the promising results from tokamaks, many alternative fusion concepts lost their support as resources became more focused on the tokamak. Many of the major tokamaks such as JET, TFTR, DIII-D, ASDEX-U, and JT-60U were built during this period. Since then, tokamaks have proved more difficult than expected, causing frustration with the pace and large-scale nature of government-funded research. This, along with the burgeoning venture capital scene, have led to a number of for-profit startup companies that are searching for a faster route to fusion power. While these companies are pursuing wildly different approaches, almost all have one thing in common: small device size (see Figure 9.9). This isn't a coincidence — smaller generally means faster development times and lower capital costs. Of course, in Chapters 7 and 8, we learned a general argument for what pushes fusion devices to large size: confinement. Because small plasmas have a large surface area

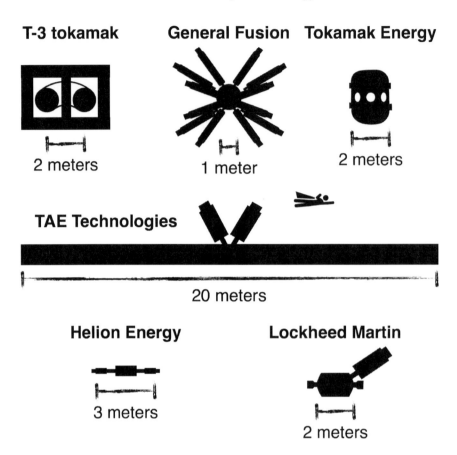

Figure 9.9: A comparison of the sizes of the T-3 tokamak and five different fusion startups devices: General Fusion's Mini-Sphere, Tokamak Energy's ST40, TAE Technologies' C-2U, Helion Energy's IPA, and Lockheed Martin's T4.

compared to their volume, their energy confinement time, and hence triple product, will tend to be lower. A confinement scheme would have to be extremely good to produce net power *and* accomplish it in a small device.

That said, nothing would be better for the fusion community (and the world), than for one of these companies to succeed. The number of fusion researchers is small and the electricity market is vast, so everyone's employment prospects would skyrocket. Thus, it would seem like the interests of government-funded fusion programs and these companies should be well-aligned. Unfortunately, there are some significant differences in their research environments. On one hand, for-profit companies must attract investors who are presumably seeking profits within a decade or so. On the other hand, government-funded research groups, especially in fusion, must maintain credibility and demonstrate steady progress. This means that the startups are more prone to make grander and less substantiated pronouncements, while the government-funded community tends to be risk adverse and less focused on producing a marketable product. This is compounded by the fact that, for financial reasons, startups tend to be more secretive and publish less.[10]

Nowhere are these differences more marked than when predicting the date of the first power plant. The startups often declare that they will build one within a decade, but then provide little evidence to show that this is possible. This undercuts the government-funded community, which is trying to maintain the funding for a device that takes two decades to build. You can see why there could be tensions. However, to be blunt, fusion within a decade is unlikely[11] and a startup being able to predict it substantially beforehand is even less likely. Plasma physics is sufficiently complex and these devices so poorly understood, that success would be more to do with luck than strategic vision. As shown in Figure 9.10, these companies are far from success. In fact, we believe they have yet to surpass the performance that the T-3 tokamak achieved in 1968. Hence, they must extrapolate their devices to plasma conditions that are a factor of 10,000 more extreme. Tokamaks are supported by a vastly larger body of research, yet can't be extrapolated by a factor of 10 without substantial uncertainty. Nevertheless, fresh ideas are always good and hold the *potential* for a quicker route to fusion. Moreover, applying the startup mentality and culture to the fusion problem

---

[10] An exception to this discussion is General Atomics, which is a large, for-profit company that was founded in the 1950s. It operates DIII-D, one of the two large tokamak experiments in the US. Because it is government-funded, General Atomics functions very similarly to a national laboratory or university.

[11] It can take that long just to get governmental approval to handle tritium.

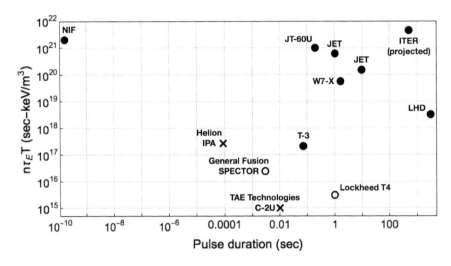

Figure 9.10: The triple products published by some of the fusion startups compared to more conventional devices [38–42]. Filled circles indicate a governmental fusion device pursuing D–T fusion, open circles indicate a startup pursuing D–T fusion, and the crosses indicate a startup pursuing an alternative fuel. Devices with shorter pulse durations and those using alternative fuels require significantly higher triple products for energy production.[12]

is worth a try. The challenge becomes balancing funding for them against the mainstream efforts. Detailed information is scarce as these fusion companies are primarily supported by private investors, but overall they have budgets significantly lower than government-funded projects. For example, TAE Technologies appears to be the most successful at fundraising. Since the company was founded in 1998, it has reportedly raised hundreds of millions of dollars and supports a staff of around 150 people. This is fairly similar in scale to a medium-sized tokamak like TCV or Alcator C-Mod.

Below we will present a short profile of some of the most notable fusion startups as well as give our own opinions on their outlook. Since fusion devices are difficult to wrap your head around, regardless of your level of expertise, we contacted many of them to better

---

[12] We could not find any published experimental results from Tokamak Energy Ltd. and they did not respond to our requests for information.

understand the state of their research as well as the credibility of their approach. These are some of the questions we asked:

(1) What fusion triple product have you achieved and what value do you need for a power plant? How do you extrapolate plasma behavior between these points?
(2) What sets the energy confinement time of the device? Do you expect turbulence to become important as the plasma temperature and density are increased?
(3) Is your plasma MHD stable?
(4) If your plasma must be compressed, do you expect to encounter instabilities (e.g. the Rayleigh–Taylor)?
(5) If you don't use deuterium–tritium fuel, why not?
(6) What power conversion efficiencies do you need for a power plant (i.e. the difference between Q and Q_{eng} in Figure 5.11)?

When you encounter a researcher working on a novel fusion concept, you should ask them some of these.[13]

9.3.1 *Tokamak Energy Ltd.*

Tokamak Energy Ltd., a British company, is one of the more conventional of the private fusion startups. Like the mainstream academic community, they are developing power plants based on D–T fusion in tokamaks. However, they hope to enable very small reactors by using high-temperature superconductors in what are called *spherical tokamaks*. Spherical tokamaks are like conventional tokamaks, except they have a very large plasma cross-section relative to the size of the device. So they look more like an apple with its core removed than a donut (see Figure 9.11). Spherical tokamaks, while less common than conventional tokamaks, have still been studied extensively. This work has revealed several advantages compared to conventional tokamaks, which likely motivate Tokamak Energy's interest. First, the kink limit becomes less restrictive in spherical

---

[13]Regardless of the setting or context. There's nothing fusion researchers like more than being grilled about MHD stability at a party/social gathering.

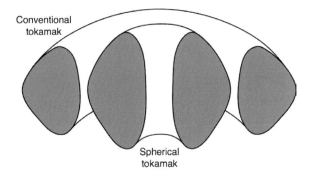

Figure 9.11: Spherical tokamaks have a larger cross-sectional radius compared to their toroidal radius.

tokamaks, allowing higher plasma currents.[14] Second, it has been shown that, by virtue of their geometry, they can directly violate the Troyon limit somewhat. Both these facts enable a device to confine more plasma pressure and achieve a higher fusion power density — a crucial parameter for power plant economics. Finally, as indicated in Figure 9.11, the plasma cross-section of spherical tokamaks can be given a stronger "D" shape without exciting MHD instability. This improves the plasma's energy confinement time, which we learned about when discussing JET's design in Chapter 6.

For these reasons, spherical tokamaks are very attractive. However, while they are useful for studying plasma physics, they are *not* normally envisioned to be power plants. The primary difficulty of a spherical tokamak power plant is fitting everything into the donut hole of the device. In a conventional tokamak, the following components take up space there:

- the vacuum vessel,
- the blanket,
- neutron shielding,
- the toroidal field coils, and
- the central solenoid.

---

[14]This can be seen from Equation (8.8) in the Tech Box on page 247.

With the possible exception of the central solenoid, you still need all of these components in a spherical tokamak. Hence, to save space you have to make sacrifices.

For example, if you reduce the amount of neutron shielding, it will reduce the lifetime of the superconducting coils and increase the external power needed to keep them cold.[15] More concerning, however, is that space constraints reduce the amount of structural support you can include in your coils. This very well could prevent Tokamak Energy from taking full advantage of the high-temperature superconductors they plan to use. Remember, in Chapter 8 we learned that there are two limits that constrain the maximum magnetic field — mechanical stresses in the coils and the magnetic field allowed by the superconductor. We only care about the one that is most limiting. The big advantage of high-temperature superconductors is that they can allow very high magnetic fields, effectively removing the second limit. However, since spherical tokamaks have little space for structural support, it seems that the mechanical limit would be more constraining anyways, so high-temperature superconductors wouldn't be as helpful.

Nevertheless, since the new high-temperature superconductors can be operated at higher temperatures, they can better handle the heating from neutrons. Moreover, they look to be a bit more resilient to neutron damage than traditional superconductors. These improvements (along with getting rid of the central solenoid) might allow a tokamak power plant to be made a bit more spherical and get a boost in performance. However, the irony is, if you constrain the donut hole to be a certain size, the only way to make a tokamak spherical is to make it very big (see Figure 9.12).

Despite viewing their plan to deliver electricity to the grid by 2030 as overly optimistic, we agree with most of Tokamak Energy's aims. Small tokamaks are desirable as they minimize

---

[15]The spherical tokamak power plant designs that do exist are just about the only ones to use copper toroidal field coils (rather than superconductors). Copper is much more resilient to neutron damage and can better tolerate the heating from neutrons.

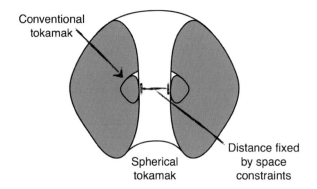

Figure 9.12: If the minimum size of the donut hole is fixed by space constraints, then spherical tokamaks must be much larger than standard tokamaks.

construction time and cost, allowing you to iterate through design ideas more quickly. Spherical tokamaks definitely deserve further research. Moreover, it is important to develop high-temperature superconductors as they present a great opportunity for fusion.[16] Lastly, finding clever ways to reduce the diameter of the donut hole, which is a significant challenge for Tokamak Energy, would be useful for all tokamaks. Despite pursuing similar goals as the government-funded community, we imagine that Tokamak Energy, as a for-profit company, may have a different way of approaching the problem. Perhaps it will pay dividends.

9.3.2 General Fusion

Of all the fusion startups, General Fusion certainly has the most visually striking design. Looking at its silhouette in Figure 9.9, you see a large number of "arms" sticking out everywhere. These are pistons that are used to compress a magnetically confined plasma. General Fusion is pursuing a small power plant with what is called *magnetized target fusion* (or MTF). Magnetized target fusion is quite literally a direct compromise between magnetic and inertial

---

[16]In fact, to our knowledge, Tokamak Energy is the first group to have built a tokamak using high-temperature superconducting magnets, although it's quite small.

Table 9.1: The desired triple product parameters for the three different confinement techniques. The parameters that General Fusion is aiming for are used as the example of magnetized target fusion.

Fusion strategy	Temperature	Density	Confinement time
Magnetic	10 keV	10^{20} m^{-3}	1 sec
Magnetized target	10 keV	10^{23} m^{-3}	0.001 sec
Inertial	10 keV	10^{31} m^{-3}	10^{-9} sec

confinement. All three strategies aim for the same temperature, but they each strike a different balance between the energy confinement time and density (see Table 9.1). MCF uses long confinement times and low densities, ICF uses short confinement times and high densities, and MTF aims in between. To do this, magnetized target fusion starts with a magnetically confined plasma and then compresses it to increase the density à la inertial confinement. The downside is that you get the negative aspects from both MCF and ICF — MTF usually needs large external magnets, a way to drive compression, and is inherently pulsed. The upside is that, by striking the right balance, you might be able to avoid the worst effects of turbulence and the Rayleigh–Taylor instability. Like most fusion concepts, MTF research stretches back to the 1950s. In fact, one of the earliest examples was James Tuck's Perhapsatron at Los Alamos, which has maintained a small MTF program ever since.

General Fusion's device, shown in Figure 9.13, is composed of a spherical tokamak vacuum vessel partially filled with a conducting liquid made of lithium and lead. Using external pumps, the liquid metal is made to rotate. Hence, like a top-loading laundry machine, the centrifugal force pushes the liquid outwards and up the vessel wall to create a vortex and cylindrical cavity. Next, a D–T plasma is injected into the cavity to form a spherical tokamak.[17] Finally, the pistons surrounding the vessel smoothly push additional liquid

---

[17]You don't have to worry about non-inductive current drive because the plasma doesn't last for long.

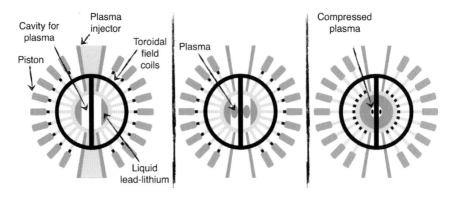

Figure 9.13: A cross-sectional view of the three stages envisioned in General Fusion's power plant: Create a spinning vortex of liquid lithium-lead, inject a spherical tokamak plasma, and use pistons to compress the lithium lead around the plasma.

metal into the chamber, collapsing the cavity around the plasma.[18] The pistons are arranged and timed in such a way as to provide a uniform and spherical compression. General Fusion believes that, if they can compress the initial plasma by a factor 300, it will ignite. The surrounding lithium-lead would absorb the resulting neutrons and breed tritium. Repeating a pulse every few seconds would provide sufficient heating to run a steam turbine and generate electricity.

The use of liquid lithium-lead is elegant as it both provides compression and serves as a blanket, minimizing the neutron damage to permanent solid components. Unfortunately though, the liquid does not shield the central column of the toroidal field coils. Hence, like all spherical tokamaks it seems challenging to ensure their survival. You could always add dedicated neutron shielding, but this would make the device much bigger. Another option that General Fusion has explored is to just remove the toroidal field coils and inject

---

[18] As General Fusion has learned more about these devices, their design has evolved. Previously, they had planned to slam the pistons into the liquid metal to drive a shock wave and achieve more extreme compression. Recently, they have transitioned to a slower, smoother compression. However, this leads to less compression overall, so the plasma must have a longer confinement time.

the plasma in anyways. These configurations are called *spheromaks* and are analogous to a smoke ring, except made out of plasma. When you blow a smoke ring, it ultimately disintegrates, but can sustain itself for a limited time. Similarly, spheromaks ultimately disintegrate, but, if the compression happens quickly enough, this might be tolerable.[19] The disadvantage is that they have significantly lower energy confinement times compared to spherical tokamaks.

One potential problem with the liquid lithium-lead blanket is that it has enormous potential to contaminate the plasma. One of the most important steps in achieving good performance on tokamaks was to fastidiously clean the first wall. This reduced the presence of heavy impurity atoms, which dilute the fusion fuel and radiate energy away via electromagnetic waves. In this way, the lead atoms in the liquid lithium-lead are concerning. Between the repeated compression of the blanket and the presence of a high-temperature plasma, it seems likely that large amounts of lead will find its way into the plasma.

Nevertheless, we applaud General Fusion's innovative approach and its openness. In contrast to many other fusion startups, General Fusion is more engaged with the broader fusion community. While we think their timeline is optimistic,[20] a fresh look at magnetized target fusion is certainly money well spent.

9.3.3 *Lockheed Martin*

Recently, there has been much hype in the press about Lockheed Martin's foray into fusion. Unfortunately, we know little about it as Lockheed is one of the most secretive of the commercial fusion ventures. While some presentations have been given [40, 43], to our knowledge there have been no peer-reviewed publications on their device, the Compact Fusion Reactor. Lockheed appears to be pursuing steady-state D–T fusion in a modified magnetic mirror. By

---

[19]Who knew that vaping could teach you about fusion?
[20]In particular, we are skeptical of the confinement time scaling they use to extrapolate to a power plant.

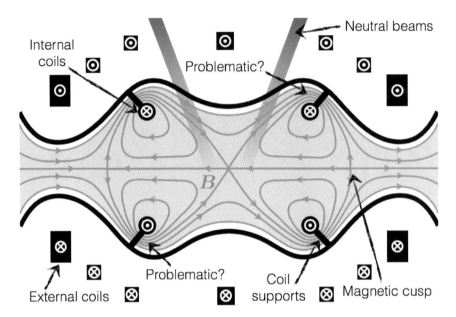

Figure 9.14: A cross-sectional view of Lockheed Martin's fusion concept. To visualize the design in three dimensions, imagine rotating this picture around the straight horizontal magnetic field line in the middle. By placing coils inside a magnetic mirror, they can create a cusp geometry. The ⊙ symbol in the coils indicates the current is directed into the page, while the ⊗ symbol indicates it comes out of the page.[21]

adding extra coils *inside* the plasma, they create what is called a *cusp geometry* (shown in Figure 9.14). They claim that doing so will ensure MHD stability at high plasma pressure and improve confinement. This isn't unreasonable as magnetic field lines that curve away from the plasma have long been known to reduce turbulence. In Lockheed Martin's design, the magnetic field lines have this good curvature everywhere, except for the locations next to the coil supports, which we've labeled in the figure as "Problematic?"

---

[21] Strictly speaking, the magnetic field shown in this figure is not accurate when the plasma is present. Lockheed plans to confine so much fuel in this device, that the motion of the particles will almost completely cancel the magnetic field in the plasma.

However, these problematic regions could be enough to doom the machine. Since they have bad magnetic field line curvature, we would expect instabilities like turbulence to be most active there. Moreover, these region are very thin, so we expect the temperature gradient between the fusion plasma and the solid material to be extremely large, which is a strong drive for turbulence. This is especially troubling because these regions are right next to the internal coils, which Lockheed plans to make superconducting [44]. All the heat that makes it to the coils (quite possibly megawatts) must be removed using external power. Finally, these internal coils don't just float in the plasma. They need solid mechanical supports, which pass through the plasma and connect to the vacuum vessel. Since Lockheed is aiming for a magnetic field strength of several Tesla, these supports will need to withstand strong magnetic forces. Additionally, all of the coolant required to maintain the superconductors at cryogenic temperatures must pass through these supports. Thus, it seems like these supports will be large.[22] Consequently, we run into the same problem we found for electrostatic confinement way back in Chapter 4 — how do you prevent the plasma from hitting the solid material?

To address this problem, Lockheed has proposed magnetically shielding the supports [45], but this seems unlikely to be sufficient. As in any magnetic fusion device, the particles are moving so fast that they will travel the distance between the Earth and the Moon before they fuse. They will cross the device *billions* of times. Even if there is an infinitesimally small chance of hitting these supports, it will ruin the confinement time of the device.

To our knowledge, Lockheed Martin has yet to present results showing that they have even produced a fully ionized plasma. Yet, they assert that a 15.5 meter × 6.5 meter device will be sufficient to produce 100 MW of electricity. And they plan to accomplish this by 2025. We are skeptical.

---

[22] Additionally, the coils themselves will need to be large. To protect the superconductors from neutrons and enable them to survive for 10 years, we believe they must be surrounded with *at least* half a meter of neutron shielding.

9.3.4 TAE Technologies

TAE Technologies, founded in 1998, is a startup based in California that is pursuing steady-state magnetic confinement fusion using proton-boron (p-B) fuel. Their concept is based on what is called a *field-reversed configuration*. Field-reversed configurations are very similar to the spheromaks that we briefly discussed earlier.[23] They are both like smoke rings made out of plasma and both require an externally created magnetic field for stability. Figure 9.15 shows the reason for the name "field-reversed configurations" — there is a current running through the plasma, which generates a magnetic field

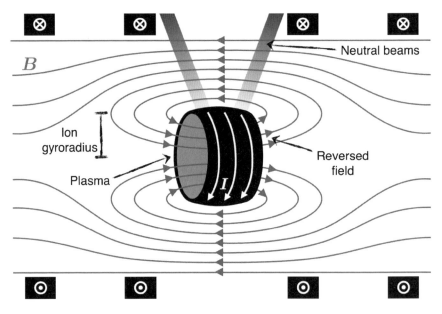

Figure 9.15: A schematic of a field-reversed configuration sustained by neutral beam current drive. If you tilt your head or the book (but not both!) by 90 degrees, you can see a tokamak-like plasma, complete with a plasma current, but no toroidal field. Note that the size of the current loop (and hence the plasma) is comparable to the particle's radius of gyration around magnetic field lines.

---

[23]The only difference is that spheromaks have both a toroidal and poloidal magnetic field, while field-reversed configurations just have a poloidal one.

that *reverses* the externally imposed one. Research on them began in the 1960s, primarily at the US Naval Research Laboratory and Los Alamos Laboratory.

To start a plasma discharge in TAE's device, you first turn on external coils in order to generate a straight magnetic field throughout a long cylindrical vacuum vessel (see Figure 9.16). Then, you "blow" in two field-reversed configurations, one from each end (i.e. from the left and right sides in Figure 9.15 or 9.16). These collide and merge to make one single hotter field-reversed configuration, which is at rest in the center of the device. At this point, unless you do something special, the field-reversed configuration will quickly lose steam and diffuse away (just like smoke rings). In order to sustain it in steady state, you must externally drive the plasma current, which TAE does using neutral beams.

Field-reversed configurations are attractive because it appears that they can make better use of their magnetic field. An external magnetic field of a given strength can stably confine more plasma in a field-reversed configuration than in a tokamak. Unfortunately, it is difficult to maintain them in steady state and their energy confinement time has always been much lower. However, since there

Figure 9.16: TAE Technologies' brand-new flagship device: Norman.

has been relatively little work on field-reversed configurations, TAE may discover ways to overcome these problems. This is certainly an effort that we support. The thing we find troubling about TAE's approach is their choice of proton-boron fuel. This choice is a dramatic break from the "mainstream" fusion community, which exclusively aims for D–T fuel.

The proposed reaction,

$$^{1}\text{H} + {}^{11}\text{B} \to 3\,{}^{4}\text{He} + 8.7\,\text{MeV}, \tag{9.1}$$

does indeed have some advantages. It uses naturally occurring, abundant fuel and the reaction is *aneutronic*. This means that it produces no neutrons, which drastically simplifies power plant design. If you use aneutronic fuel, your device no longer requires neutron shielding or a lithium blanket. Moreover, instead of a steam cycle, you could try generating electricity directly from the motion of the charged particles. This may enable higher conversion efficiencies. Hence, proton-boron fusion is the "ideal" fuel source.

That said, as we saw in Chapters 3 and 4, there are good reasons to use D–T fuel, especially given that TAE's device appears to have space for neutron shielding. The D–T fusion cross-section is a factor of 10 larger than for p-B and this peak occurs at a temperature 10 times lower. The lower temperature makes plasma confinement easier and the higher cross-section means that the fusion power density of a D–T reactor would be 10 times larger. The advantage of D–T is clear from Figure 9.17, which shows the triple product required for ignition with different fuels. Furthermore, this figure does not capture the most challenging downside of proton-boron fusion: particle radiation.

When discussing the energy confinement time in Chapter 4, we mentioned that particles in the plasma radiate electromagnetic waves, which carry energy out of the plasma. This radiation is a fundamental physical effect that occurs whenever a charged particle accelerates, decelerates, or changes direction. However, we neglected it, saying that the energy carried by electromagnetic waves was small compared to the energy losses due to turbulence. In p-B plasmas, this is no longer true.

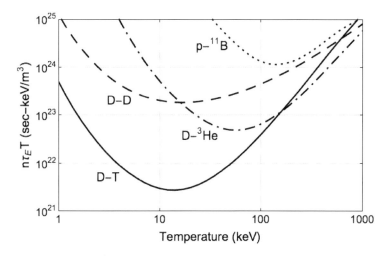

Figure 9.17: The triple product needed to ignite proton-boron fuel compared to the other fusion reactions that we have discussed.

The most inescapable radiation is called *bremsstrahlung*, which is German for "braking radiation." Bremsstrahlung refers to the radiation that a charged particle emits when it collides with another particle. Hence, we see that it is inextricably tied to the process of fusion. In order to get two particles to fuse, they must collide, but the vast majority of the time they simply bounce off one another and emit bremsstrahlung without fusing. This process emits more energy if the particle is moving faster, so electrons radiate much more than the ions.[24] Herein lies the problem.

As seen in Figure 9.17, proton-boron fusion requires higher temperatures than the conventional fuels, which increases the amount of bremsstrahlung. However, a bigger effect comes from the fact that a boron atom has five electrons. Therefore, when you add it to a plasma, you are adding five free electrons (all of which

---

[24]Since the electrons are only useful to ensure quasineutrality (because they don't fuse), you would like them to be at a much lower temperature than the ions. However, because collisions are so common and tend to equalize the two temperatures, this can only be done to a limited extent (see the "Thermal equilibrium" Tech Box on page 82).

produce bremsstrahlung), but only one nucleus that can undergo fusion. Accounting for all this mathematically, you find that, even for optimal plasma conditions, the bremsstrahlung looks to radiate away more power than p-B fusion will produce (see the following Tech Box). This kills the energy confinement time and prevents ignition, making a p-B power plant appear nigh impossible.[25] The only way to make a power plant produce net energy is to have really high electric conversion efficiencies (i.e. fusion power to electricity and electricity to plasma heating), much higher than current technology appears to allow. Thus, we see p-B fusion as an idealistic pursuit, rather than a practical way to put watts on the grid. And as fusion researchers, that's saying something.

TECH BOX: Bremsstrahlung and proton-boron fusion

Understanding the impact of bremsstrahlung requires considerable mathematics, so you should feel free to skip this Tech Box. To those of you that remain, the fusion power produced per unit volume in a p-B plasma is given by

$$p_{\text{fusion}} = n_p n_B \langle \sigma v \rangle E_{\text{fusion}}, \quad (9.2)$$

where $E_{\text{fusion}} = 8.7\,\text{MeV}$ is the energy released by each fusion reaction and n_p and n_B are the number densities of protons and boron nuclei respectively. The quantity $\langle \sigma v \rangle$ (which was discussed in the Tech Box on page 132) is an average over the fusion cross-section and depends on the ion temperature T_i. Because of the high atomic number of boron, it is best to use a 5:1 ratio of protons to boron nuclei to maximize the fusion power. We are interested in comparing this fusion power to the bremsstrahlung power produced per unit volume of plasma,

---

(*Continued*)

---

[25] A more general and rigorous formulation of this argument can be found in Todd Rider's 1995 MIT PhD thesis [46].

(*Continued*)

which is given by [46, 47]

$$p_{\text{brem}} = \frac{160}{3} \left(\frac{e^2}{4\pi\epsilon_0}\right)^3 \frac{n_e^2 \sqrt{k_B T_e}}{(m_e c^2)^{3/2} \hbar}$$

$$\times \left(\frac{Z_p^2 n_p + Z_B^2 n_B}{n_e}\left(1 + .79\frac{k_B T_e}{m_e c^2} + 1.9\left(\frac{k_B T_e}{m_e c^2}\right)^2\right)\right.$$

$$\left. + \frac{3}{\sqrt{2}}\frac{k_B T_e}{m_e c^2}\right). \tag{9.3}$$

Here, $e = 1.60 \times 10^{-19}$ C is the electric charge of a proton, $\epsilon_0 = 8.85 \times 10^{-12}$ F/m is the vacuum permittivity, $k_B = 8.62 \times 10^{-8}$ keV/K is the Boltzmann constant, $m_e c^2 = 511$ keV is the electron rest energy, $\hbar = 6.58 \times 10^{-19}$ keV-sec is the reduced Planck constant, $Z_p = 1$ is the atomic number of hydrogen, $Z_B = 5$ is the atomic number of boron, n_e is the number density of electrons, and T_e is the electron temperature. After crunching the numbers for roughly optimal conditions [46], we find this table:

Fuel	$k_B T_i$	$k_B T_e$	$\langle \sigma v \rangle$	$p_{\text{fusion}}/p_{\text{brem}}$
D–T	50 keV	42 keV	8.7×10^{-22} m^3/s	139
D–^{3}He	100 keV	73 keV	1.7×10^{-23} m^3/s	6.6
D–D	500 keV	209 keV	2.0×10^{-22} m^3/s	2.9
p–^{11}B	300 keV	137 keV	2.4×10^{-23} m^3/s	0.57

We see that, in contrast to the conventional fuels, a p-B plasma radiates away substantially more power than it creates from fusion, so it appears impossible to ignite. Nevertheless, if the power plant technology was efficient enough, it could still produce net energy. To see how efficient, we will look

(*Continued*)

(*Continued*)

at Figure 8.1 (and the footnote therein) and write the net electrical power density as

$$p_{\text{net}} = \eta_{\text{steam}} \left(p_{\text{fusion}} + p_{\text{heat}} \right) - \frac{p_{\text{heat}}}{\eta_{\text{h\&cd}}}. \qquad (9.4)$$

Here, η_{steam} is the efficiency by which the energy exhausted by the plasma can be turned into electricity, $\eta_{\text{h\&cd}}$ is the efficiency by which the heating systems convert electricity into thermal energy in the plasma, and p_{heat} is the heating power density required to keep the plasma hot. To be as generous as possible, we will assume that bremsstrahlung is the only energy loss mechanism (i.e. no turbulence), so that $p_{\text{heat}} = p_{\text{brem}} - p_{\text{fusion}}$. For a p-B plant to produce any net energy at all, we require that $p_{\text{net}} > 0$. Plugging in Equation (9.4) and the expression for p_{heat} gives

$$\eta_{\text{steam}} \eta_{\text{h\&cd}} > 1 - \frac{p_{\text{fusion}}}{p_{\text{brem}}}. \qquad (9.5)$$

This is a very difficult requirement to meet. Even a p-B plasma that attained the incredible temperature of 300 keV (and did so without exciting any turbulence) would still have $p_{\text{fusion}}/p_{\text{brem}} = 0.57$, which requires a total recirculating power efficiency of $\eta_{\text{steam}} \eta_{\text{h\&cd}} > 0.43$. In Chapter 5, we learned that the efficiency of neutral beams alone is typically below this, with $\eta_{\text{h\&cd}} = 0.3$. Moreover, meeting this condition only gives you net electricity. Producing enough for economic viability is harder.

9.3.5 *Lawrenceville Plasma Physics*

Lawrenceville Plasma Physics (LPP) is a small fusion startup located on the outskirts of Newark, New Jersey.[26] They are pursuing a

---

[26] In 1991, the founder of LPP wrote a book titled *The Big Bang Never Happened: A Startling Refutation of the Dominant Theory of the Origin of the Universe*. Unsurprisingly, it hasn't persuaded the astrophysical community.

strategy called *dense plasma focus*, which involves connecting a bank of capacitors to two electrodes with a gas of fusion fuel in between. Just like lightning or a Tesla coil, this will cause an arc of electricity to jump across the gap and ionize the fusion fuel. Since parallel electric currents attract one another, the arc of electricity pinches together, compressing and heating the fuel — ideally to fusion conditions. Unfortunately, the arc is plagued by plasma instabilities, such as the tendency to kink that we learned about when discussing MHD stability, so it disintegrates quickly. Dense plasma focus devices have been studied since the 1950s and, while potentially useful as a source of X-rays or neutrons, are generally not envisioned to be workable power plants.

LPP's device achieves high energies, but is still far from net fusion power. According to a recent publication [48], it has achieved ion energies of 250 keV. However, each discharge lasts for just 40 nanoseconds, which is much shorter than the ion-ion collision time of a microsecond. This means that the ions in LPP's device do not undergo enough collisions to be in thermal equilibrium. Therefore, it's not appropriate to assign a temperature to them (or a triple product). In fact, only about 1 in every 25 ions even collides with another ion, let alone fuses. In effect, it looks like the device is acting as a particle accelerator, creating a beam of energetic ions that collide with a background thermal plasma to produce some fusion reactions (i.e. similar to the ZETA results discussed on page 165). This behavior is in keeping with established theoretical understanding as well as other dense plasma focus experiments [49]. Unfortunately, as we saw in Chapter 3, accelerators are difficult to make into power plants. To have a viable device, LPP must make fundamental improvements to the dense plasma focus concept. They must somehow significantly lengthen the pulse duration (i.e. improve confinement) and/or increase the density in order to enable thermal equilibration. After that, further progress would be required to achieve enough fusion reactions for net power production.

Their path is made much harder because, like TAE, LPP plans to use proton-boron fuel. This appears necessary because the solid electrodes are very close to the plasma and would have trouble withstanding the neutron bombardment from D–T reactions. Thus,

LPP is also confronted with the problem of electron bremsstrahlung. To overcome it, LPP has proposed using what they call the *quantum magnetic field effect* (see the following Tech Box) [50]. The idea is that, if *extremely* strong magnetic fields can be created by the device, quantum mechanics will start to modify the gyrating motion of the electrons. Theoretically, this might be able to reduce the energy that collisions transfer from the ions to the electrons. If this could be accomplished, it might be possible to maintain the ions at the high temperatures needed for p-B fusion, while the electrons remain cold enough to make bremsstrahlung radiation tolerable.

TECH BOX: Quantum magnetic field effect

According to quantum mechanics, the gyration of charged particles around a magnetic field line is quantized. That is to say that particles can only have specific, discrete velocities, which correspond to what are called *Landau energy levels*. These energy levels are evenly spaced at

$$\Delta E = \hbar \frac{qB}{m}, \tag{9.6}$$

where $\hbar$ is the reduced Planck constant, q is the electric charge of the gyrating particle, B is the magnetic field strength, and m is the mass of the gyrating particle.

Normally, this discretization is impossible to notice because the levels are so close together that everything appears smooth and continuous. For example, the strongest magnetic fields in current tokamaks are around $B = 10$ Tesla, which lead to electron energy levels that are 0.000001 keV apart. This spacing is tiny compared to typical fusion plasma temperatures of 10 keV. However, if the magnetic field becomes very, very strong, then the spacing between the electron energy levels can become comparable to the plasma temperature. If this were to happen, the quantum magnetic field effect could begin to modify the thermal equilibration of the plasma.

(Continued)

(*Continued*)

The most important modification concerns the energy transfer from the ions to the electrons. Because of the fusion cross-section, achieving substantial p-B fusion requires ion temperatures of roughly $k_B T_i = 300\,\text{keV}$. Normally, since collisions are so common, this means that the electrons must also be about 300 keV, leading them to radiate unacceptable amounts of energy via bremsstrahlung. However, if the spacing between electron energy levels is larger than the amount of energy that can be transferred in a single ion-electron collision, then ion-electron collisions will stop occurring. This threshold for collisions is similar to what enables superconductivity (see page 183). The condition to prevent ion-electron collisions is given by

$$\Delta E > \frac{m_e}{m_i} k_B T_i, \tag{9.7}$$

where ΔE is the electron energy level spacing from the previous formula and m_e and m_i are the masses of the electrons and ions, respectively. The reason for the mass ratio factor is that, when particles have different masses, they can't transfer all of their energy in a single collision. Think of playing billiards versus bowling. Two billiard balls have the same mass, so, when the cue ball hits the 8 ball, it can transfer all of its energy and wind up stationary. Bowling balls, on the other hand, are much more massive than bowling pins, so they plough through them and only transfer a fraction of their kinetic energy. Because ions are so heavy, they plough through the electrons like a bowling ball and only transfer a small amount of their energy.

If the magnetic field is strong enough to meet Equation (9.7), then most ions won't be able to transfer their energy to electrons because the gaps in the allowable electron energy levels would be too large. For 300 keV ions, this requires a magnetic field strength of more than a million Tesla.

While limiting electron bremsstrahlung using the quantum magnetic field effect is a neat theoretical idea, this phenomenon has never been demonstrated experimentally. We are skeptical that any fusion device would be able to achieve the required magnetic field strength. LPP is aiming for a field of a million Tesla, which is a thousand times stronger than the strongest field ever created by non-destructive means (i.e. without destroying the device that created it). It is on par with the magnetic field near a neutron star. Unless these incredible fields can be created (and in an energy efficient manner), p-B fusion still appears doomed by the energy losses from electron bremsstrahlung.

9.3.6 *Helion Energy*

Helion Energy, a startup based in Redmond, Washington, is working on a device they call "the Fusion Engine." The fusion engine is a magnetized target fusion device that aims to fuse deuterium and helium-3 fuel via the reaction

$$^{2}\text{H} + {}^{3}\text{He} \rightarrow {}^{4}\text{He} + {}^{1}\text{H} + 18.3\,\text{MeV}. \tag{9.8}$$

To achieve this, they plan to use field-reversed configurations and compression, so it's kind of like a combination between the approaches of General Fusion and TAE. Helion's idea is to create two field-reversed configurations at either end of a long vacuum vessel and accelerate them towards one another using external magnetic fields. Then, at the moment of collision, they rapidly increase the strength of the external magnetic field to magnetically compress the plasma and achieve substantial fusion. Like TAE, they hope to directly convert the kinetic energy of the D–^{3}He fusion products into electricity (instead of using a thermal steam cycle). This technology is still very much under development, but theoretically could enable higher conversion efficiencies.

What's most puzzling about Helion's strategy is their use of D–^{3}He fuel. We certainly agree that the D–^{3}He reaction is attractive. It is fairly aneutronic[27] and the triple product needed for its ignition is second only to D–T (see Figure 9.17). In fact, Helion's website

---

[27]The D–^{3}He reaction does not produce any neutrons directly. However, D–D reactions invariably occur in the plasma and produce some neutrons.

notes that "D–^{3}He fusion simplifies the engineering of a fusion power plant, lowers costs, and is even cleaner than traditional fusion." The catch is that helium-3 doesn't exist on Earth in sufficient quantity. So, how does Helion obtain it? Well, according to one of their patents [51], the plan is to obtain it from the D–D fusion reaction

$$2\,^2\text{H} \begin{matrix} \nearrow\,^3\text{H} + {}^1\text{H} + 4.0\,\text{MeV} \\ \searrow\,^3\text{He} + {}^1\text{n} + 3.3\,\text{MeV}. \end{matrix} \qquad (9.9)$$

Unfortunately, unlike D–^{3}He fusion, D–D reactions produce neutrons and tritium,[28] which complicates the engineering of a fusion power plant, increases costs, and leads to radioactive waste. Moreover, the D–D reaction is harder than D–^{3}He as it requires a higher triple product (see Figure 9.17). At best, Helion's marketing is misleading — they tout the benefits and ease of aneutronic D–^{3}He fusion, but still need the harder neutronic D–D fusion to obtain their fuel.[29]

9.3.7 *Commonwealth Fusion Systems*

Lastly, we must mention a very recent upstart in the world of startups. In March 2018, a brand-new private company, Commonwealth Fusion Systems, announced that it had received an investment of $50 million and is eagerly seeking additional support. Commonwealth was founded by a group of academic fusion scientists, primarily researchers who worked on the recently-shuttered Alcator C-Mod tokamak at MIT. They aim to build a compact D–T tokamak experiment called SPARC using high-temperature superconductors.[30]

---

[28]Tritium will undergo radioactive decay into helium-3 with a half-life of 12 years.
[29]We reached out to Helion, but they did not respond to our inquiries. We should mention that helium-3 could potentially be mined from the Moon, but Helion claims their helium-3 is "self-supplied."
[30]SPARC is a nested acronym. It stands for Smallest Possible ARC, while ARC stands for Affordable, Robust, and Compact. Acronym-ception! ARC also happens to be the name of the fusion reactor in the Iron Man comics. Coincidence? No. One of us was part of the ARC design team and was present at its naming.

SPARC is envisioned to generate 100 MW of fusion power for 10 seconds at a time and illuminate a path to a demonstration power plant. In essence, you can think of it as a less ambitious ITER. However, instead of enabling this performance by going to a large device, Commonwealth hopes to enable it using high magnetic fields produced by high-temperature superconductors. In Chapter 8, we learned why this is a promising approach that could reduce a power plant's construction costs.

Of course, trying anything new in fusion involves considerable uncertainty. If the magnetic field is significantly higher than any existing tokamak, then scientists are less confident in predicting plasma behavior. Simple extrapolations from existing tokamaks predict that the confinement time should increase substantially, but by how much? Will it be difficult to get the plasma into H-mode? To what degree will the MHD limits we discussed in Chapter 8 still hold? Finally, will the power exhausted by the plasma come out in such a small area that it melts the divertor plate?

While these are all fascinating questions, Commonwealth must first build the magnets before they explore them — no small task. High-temperature superconductors are still cutting-edge technology and the large coils needed for fusion have yet to be developed (though Tokamak Energy has started work on it).[31] Thus, most of Commonwealth's initial investment will probably be used to demonstrate the practicality of large, high-temperature superconducting magnets. Nevertheless, SPARC would complement ITER nicely. The high magnetic field route and the large size route to a power plant are not distinct, mutually exclusive options. Comparing the successes and limitations of the two machines would help determine the optimal balance and lead to a better demonstration power plant.

While Commonwealth has set quite an optimistic goal — a demonstration power plant within 15 years — we're very excited to see what they can do.

---

[31]Commonwealth and Tokamak Energy are pursuing a similar strategy. The primary difference is that Tokamak Energy wants to use spherical tokamaks, rather than conventional ones.

Chapter 10

Fusion and Nuclear Proliferation

At 5:29 am on July 16, 1945, the Jornada del Muerto desert in New Mexico saw one of the most consequential acts in human history. A bomb was detonated by spherically imploding two hemispheres of plutonium, each weighing 3 kilograms. This bomb, however, was different from all bombs before it. Instead of using the energy stored in chemical bonds, the "Trinity" test, as it is known, unleashed the energy stored in nuclear bonds. About 7 grams of plutonium disappeared instantaneously, generating the energy of more than 20,000 tons of TNT[1] (see the "Yield of a nuclear weapon" Tech Box). While physicists and chemists had theorized about the possibility of a runaway fission reaction for over a decade, the Trinity test marked a discontinuity in human history — a transition from one state to another. No longer would humans be limited by the energy available in chemical bonds. We had discovered nuclear reactions that produced millions of times more energy.

TECH BOX: Yield of a nuclear weapon

The yield of a nuclear weapon, which is the amount of energy released during its detonation, is measured in tons of TNT

(*Continued*)

---

[1] Not all of the plutonium in the bomb was fissioned — much of it was scattered over the Jornada del Meurto landscape.

(*Continued*)

> equivalent. In other words, how much TNT would be required to create as powerful of an explosion? In conventional units of energy, detonating one ton of TNT is equivalent to 4.2×10^9 Joules. In unconventional units, detonating one ton of TNT is equivalent to the energy of one New York-to-Chicago road trip. Tsar Bomba, the largest ever nuclear weapon test, released 50 million tons (or 50 megatons). The energy released by all the conventional explosives during World War II was "just" 3 megatons.

As human civilization's transition into the atomic age accelerated, we created weapons that could level entire cities. On November 1, 1952, the first hydrogen bomb, Ivy Mike, was detonated, releasing 500 times as much energy as the Trinity test seven years earlier. Although Ivy Mike was intended as a conservative bomb design, it released roughly three times as much energy as all of the chemical explosives detonated throughout the entirety of World War II. Humanity now had the capacity to release several World War II's worth of energy ... in an instant.

So how carefully has humanity been managing this knowledge of nuclear physics? As of 2018, nine countries have nuclear arsenals.[2] While a nuclear weapon has not been used since World War II, there have been many incidents when nuclear war appeared imminent or nuclear accidents nearly occurred (which could have been easily interpreted as acts of aggression). The closest we came to nuclear war may have been on October 27, 1962 — the height of the Cuban Missile Crisis. Earlier that day, a U.S. U-2F spy plane had been shot down over Cuba, killing the pilot. In response, the

---

[2] The countries with nuclear weapons are Russia (approximately 7,500 warheads), the United States (7,200), France (300), China (250), Britain (215), Pakistan (100), India (100), Israel (80), and North Korea (10). During the Cold War, the total number of nuclear weapons peaked at roughly 65,000, with the USRR possessing 40,000 and the United States possessing 23,000.

Trinity. July 16, 1945. The first ever nuclear weapons test, with a yield of 20 kilotons (of TNT). Image shows the initial fireball, 16 millisecond after detonation. The highest point in the image is 200 meters above the ground.

The Enola Gay and her crew. August 6, 1945. The Enola Gay, a Boeing B-29 Superfortress, was the first plane to drop an atomic bomb onto a target, resulting in the deaths of 90,000–146,000 people.

Fat Man. August 9, 1945. The mushroom cloud over Nagasaki rose to its peak height of 20 kilometers, roughly 10 minutes after detonation.

Ivy Mike. November 1, 1952. The first ever hydrogen bomb test had a yield of 10 megatons, 500 times greater than the first nuclear weapons test, Trinity, seven years earlier.

Figure 10.1: Some photographs from the early days of nuclear weapons.

U.S.S. Beale, an American destroyer, dropped depth charges on a Soviet submarine, the B-59, as warning shots to make it surface. Unbeknownst to the Americans, the B-59 was armed with nuclear torpedoes! Under the joint control of Captain Valentin Savitsky, political officer Ivan Maslennikov, and naval officer Vasili Arkhipov, the submarine required the unanimous approval of all three before commencing any attack. Savitsky, assuming that World War III had begun, urged the firing of a 10 kiloton nuclear torpedo. Arkhipov alone vetoed the nuclear strike and urged patience. He argued that,

Figure 10.2: "Today, every inhabitant of this planet must contemplate the day when this planet may no longer be habitable. Every man, woman, and child lives under a nuclear Sword of Damocles, hanging by the slenderest of threads, capable of being cut at any moment by accident or miscalculation or by madness." John F. Kennedy addressed the United Nations General Assembly in September 1961. The Sword of Damocles is suspended by a single hair of a horse's tail, ready to fall at the smallest mistake. This allusion is frequently used to describe the perilous state of nuclear affairs, where the slightest mishap could cause untold destruction.

since they were so far submerged and could not undergo radio communication, they had no proof that war had broken out. Had Arkhipov authorized the launch, it seems likely that all-out nuclear war between the West and the Soviets would have occurred. This was quite possibly the most perilous day in human history.

The incident with the B-59 is far from the only example of a close call. In 1961, a U.S. bomber carrying two nuclear weapons broke-up over North Carolina. After the bombs were recovered, one of them was found to have been armed to detonate! The only thing that

(a) The Soviet nuclear submarine B-59, which was forced to the surface by U.S. depth charges on October 27, 1962.

(b) Vasili Arkhipov, the Soviet naval officer who opposed the firing of a nuclear torpedo from the B-59. Without his opposition, there likely would have been all-out nuclear war.

(c) A U-2F spy plane (below), which was the type that was shot down over Cuba.

(d) A Soviet SS-4 ballistic missile in Moscow, some of which were shipped to Cuba.

Figure 10.3: Some photographs from the Cold War.

had prevented its explosion was a single low-voltage electrical safety switch. In 1995, Norwegian scientists launched a rocket to study the Aurora Borealis, which Russian radar mistook for an incoming U.S. missile. For 10 minutes the Russian president Boris Yeltsin considered launching nuclear retaliation.[3] In the past two decades, dozens of nuclear warheads have gone missing and there have been at least 25 cases of nuclear explosives being stolen.

However, you may take issue with this one-sided view of nuclear weapons. Many would argue that their presence has facilitated a

---

[3]The scientists had notified Russia of the launch beforehand, but the message hadn't been relayed to the radar command.

period of unprecedented peace, whereby acts of aggression between two nuclear powers are strongly discouraged. In our opinion, while this may have been valid during the Cold War, the nature of the nuclear threat has evolved substantially since then. In the current era, nuclear weapons are far less useful for deterrence. Non-state actors, such as terrorists, are often unperturbed by the threat of mutually assured destruction. For example, there is evidence that Al-Qaeda has been actively trying to obtain nuclear material in order to set off a nuclear device in a city. Additionally, now that so many states have nuclear weapons, it has become more likely that an attack will occur, even if it isn't a rational or strategically wise action. For example, tensions between less powerful nuclear states, most notably India and Pakistan, could still cause unprecedented death and destruction.

Scary as it is, it is worth spending some time contemplating the actual consequences of a nuclear war. Simply put, nuclear weapons provide us with the tools to destroy modern civilization through nuclear holocaust. Even the most tepid nuclear conflict, say the detonation of 100 Trinity-scale warheads, could still cause significant global cooling. This is known as *nuclear winter*. Nuclear winter is caused by soot reducing the amount of sunlight that reaches the Earth's surface — the vast majority of which is literally the ashes of the cities hit by the nukes.[4] Estimates indicate that this would lead to the deaths of hundreds of millions to billions of people. A total nuclear war between the United States and Russia would almost certainly result in the majority of humanity dying within a year, although the total extinction of the human species appears much less likely.[5] While there is some uncertainty in the scientific models used for calculating the effects of nuclear winter, when a large number

---

[4]This might be the most depressing sentence possible.

[5]The total extinction of the human species is extremely hard to accomplish by human means. There are people in far flung regions of the world, in ships on the oceans, and, in the event of war, in nuclear bunkers. Even if you set out with the world's current stockpile of 15,000 nuclear weapons and the express goal of entirely extinguishing the human species, it doesn't look to be a trivial task.

of peer-reviewed results point to civilization-terminating effects, it would be absurd not to minimize the risk. If the more damning research is correct, the next conflict in which nuclear weapons are used may well be our last. Such a conflict could set human civilization back for centuries, millennia, or even forever.

Unfortunately, nuclear weapons aren't just unique for their destructive power, they are also unique for lacking effective countermeasures. In the history of human warfare, the development of new military technology has often been reactionary. When an adversary devised a new weapon, the response would be a new form of defense to reduce its effectiveness. Stronger armor came in response to sharper blades, longer spears in response to mounted riders, slanted tank armor in response to more powerful armor-piercing shells, and high-altitude flight in response to flak and radar. Unfortunately, with respect to nuclear weapons, particularly those delivered by intercontinental ballistic missiles (ICBMs), human ingenuity has largely failed. There is no reliable defense against them, except to avoid nuclear confrontation altogether.[6]

A telling example is provided by anti-ballistic missiles (ABM), one of the leading technologies to counter the threat of ICBMs. ABMs are surface-to-air missiles designed to intercept and destroy ICBMs. The challenge is that ICBMs move very, very quickly. Upon reentry into the lower atmosphere, ICBMs can be traveling at over 25,000 kilometers per hour. The variety of methods attempted to intercept them include equipping the ABM itself with a nuclear warhead or directly intercepting it with a solid projectile. The problem with all existing ABM systems is that they don't really work. Moreover, their existence actually aggravates the situation. As a basic example, let's assume that a single ABM has a 50% chance of successfully intercepting an incoming ICBM.[7] Obviously 50% is unacceptably low, so you would fire as many ABM missiles as you can spare. If you fired 10 ABMs, then you would have a 99.9%

---

[6]It appears. Of course, the current state of this technology is highly classified.
[7]A rough estimate for today's state-of-the-art systems.

chance of success.[8] These are much better odds for you. However, your adversary knows this. Hence, it is logical for them to design ICBMs that are even harder to intercept and to fire many more of them! This means that you need to build even more ABMs in order to counter the larger number of ICBMs. And in response, your enemy will build even more ICBMs. And so on, and so forth. It is not clear where this madness ends, other than in inflated military budgets.[9]

The conclusion to all of this is that there are currently no effective nuclear defenses — there are no nuclear shields, only nuclear arrows. And it's hard to see a credible nuclear defense system coming forth in the foreseeable future. Given that there is no effective defense against nuclear weapons, and that they could easily be triggered by mistake, the logical conclusion is to minimize the number of nuclear weapons. For these reasons, it is a mainstream and credible view that a gradual, multilateral reduction in nuclear weapons, ending with complete disarmament, is not only economically and politically desirable, but likely necessary for the continuation of modern civilization. One of the surest ways to do this is to carefully regulate the material needed to make nuclear weapons: fissile material. Obtaining fissile material is the most challenging step in building a bomb. Even in a world without nuclear weapons, the existence of fissile material will make it a fragile equilibrium, a moment away from nuclear conflict.

10.1 Nuclear Physics: A Double-edged Sword

Picture a dispassionate intelligence, peering down at Earth. It would undoubtedly be fascinated, and disappointed, by how humanity has

---

[8]This assumes the probability of success for each ABM is independent, which is likely *not* a good assumption.
[9]Nuclear weapons don't come cheap. Globally, mankind spends at least $100 billion *per year* on nuclear weapons. You could do a lot with that. The Apollo space program, which ran from 1961 to 1972 cost just $10 billion per year (adjusted to 2017 dollars). The cost of meeting all of the U.N. Millennium Development Goals (eight ambitious tasks including eradicating extreme poverty, combating HIV, and achieving universal primary education) is estimated to be about $100 billion per year. And the *total* price tag for ITER is just $25 billion.

used its knowledge of nuclear physics. On one hand, we have just seen that nuclear weapons pose a continued existential risk to humanity. But on the other hand, nuclear energy and nuclear medicine undoubtedly improve human welfare. Even while the world's powers descended into the Cold War and developed increasingly destructive bombs, many had also been exploring whether controlled nuclear reactions may profit peaceful purposes. On December 2, 1942, the world's first artificial nuclear reactor,[10] Chicago Pile-1 (CP-1), was completed under one of the University of Chicago's football fields. While the motivation for CP-1 was to confirm the viability of nuclear fission for weapons applications, it laid the groundwork for a more peaceful application. Twelve years later, in 1954, the USSR connected the first nuclear fission power plant, the Obninsk Nuclear Power Plant, to the electrical grid. Since then, there has been a massive expansion of nuclear fission electricity generation. Some thirty countries now have nuclear fission power facilities, providing over a tenth of global electricity.

Clearly, nuclear physics is a double-edged sword that cuts cleanly both ways — it has the power to end our energy problems and the power to end our civilization. In 1945, the nuclear sword mainly cut militarily. In 2017, its use is much more balanced. As our mastery of fusion energy systems improves, the balance could tip even further. Regardless, the history of humanity's relationship with science is fraught with complexity. From agriculture to germ theory, computing to quantum mechanics, the scientific method does not filter for benevolence. Rather, human nature shapes knowledge into a wide spectrum of applications. Science may set the paint palette, the canvas texture, the brush thickness, but humans will determine how and what to paint. While nuclear physics is not the only field of knowledge with a dark side, its applications are among the most

---

[10]Billions of years ago, there were natural fission reactors on Earth. These occurred when water mixed with concentrated uranium deposits. Only one such natural fission reactor site has been discovered, which was in Oklo, Gabon. It appears unlikely that any natural reactors are currently operating because the natural isotopic composition of uranium has changed substantially since then.

consequential. The nuclear fission dilemma is one of the clearest manifestations of the conflicts inherent in scientific knowledge. While fission has outstanding clean energy credentials, its impact on global security gives pause. Are we willing to tolerate the inherent risk of making it a serious long-term energy provider?

In this chapter, we will investigate the impact of fission and fusion electricity generation on nuclear security. We will find that, because fission requires fissile material, fission power plants pose an intrinsic proliferation concern. Unfortunately, forgoing fission power without replacing it with low-carbon sources would aggravate an already desperate climate situation. However, fusion has the capacity to change all of this. Since fusion has no use for fissile material, the advent of fusion power plants would not only be able to revolutionize our energy system, but also improve nuclear security. By replacing fission plants with fusion ones, we remove a powerful alibi that countries can and do use to justify generating fissile material. In a fusion world, fissile material has no place but in nuclear weapons.[11] In this way, fusion power plants could facilitate nuclear disarmament and enable a world entirely free of fissile material. To see why, we will first learn more about nuclear weapons and then fission power plants.

10.2 Building Nukes

Fundamentally, nuclear fission produces energy by using a neutron to split a large nucleus into two smaller pieces and a few neutrons (see Figure 10.4). For example, uranium-235 undergoes the reaction

$$^{235}\text{U} + {}^{1}\text{n} \rightarrow A + B + 2.4\ {}^{1}\text{n} + 200\,\text{MeV}, \qquad (10.1)$$

where A and B are referred to as fission products and can be any of a number of isotopes. To generate sustained energy, fission devices

---

[11]The value of this is best illustrated by the Iranian nuclear program. The program was started in the 1950s with the purpose of electricity generation. However, in the 2000s it became unclear if Iran was using their nuclear fission infrastructure to also develop weapons. This was the problem that prompted the 2015 Iran nuclear deal.

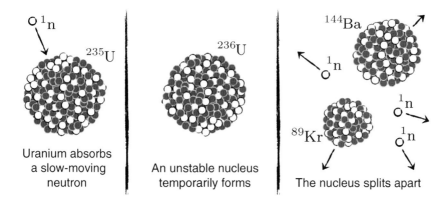

Figure 10.4: A cartoon of a uranium-235 fission event in three frames. Absorbing a neutron causes the uranium nucleus to split into two large nuclei (here barium-144 and krypton-89), release several neutrons (on average 2.4), and produce a lot of kinetic energy (on average about 200 MeV).

make use of a chain of fission reactions (see Figure 10.5). Neutrons enable this process because they both cause fission events and are produced by fission.

Finding a material that can sustain such a chain reaction is the key to fission.[12] From common sense and what we have learned, we know that any viable fission fuel must possess the following properties:

(1) It must release energy when fissioned. All nuclei with more than roughly 120 nucleons satisfy this (see the binding energy curve on page 72).
(2) It must be *relatively* stable. We want to be able to control when it fissions, so it must not decay quickly or spontaneously fission.
(3) It must be *fissile*, meaning that low energy neutrons can still cause fission.
(4) The fission cross-section must be large to enable high power density (see Figure 10.6). A lower cross-section means that

---

[12]This really can't be overstated. For fission, materials are everything.

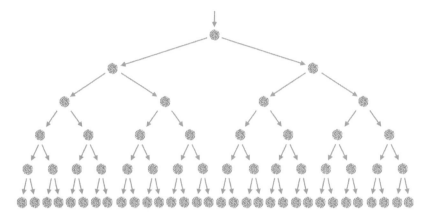

Figure 10.5: A chain of fission reactions. When a neutron (gray arrow) hits a uranium nucleus, it fissions and releases several neutrons that can each cause a successive fission event. In this example, we see that the reaction is quickly spiraling out of control, as indicated by the number of fission events in each generation.

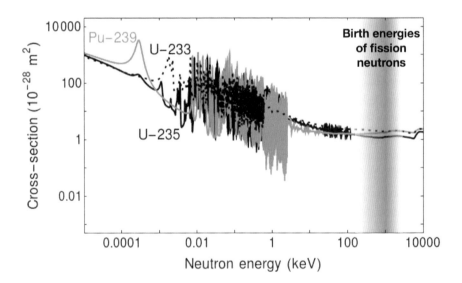

Figure 10.6: The dependence of the fission cross-section (the probability of a neutron being absorbed by the nucleus and causing a fission event) on the energy of the incident neutron.

neutrons will have to travel further to fission, so the device will have to be bigger.
(5) It must be affordable to obtain on Earth.

Only three isotopes satisfy the first four points: uranium-233, uranium-235, and plutonium-239. We will examine each of these in turn, but we will start with uranium-235 because the other two are nearly nonexistent on Earth.

Uranium is fairly plentiful in the Earth's crust, about 500 times more abundant than gold. Thus, it has been the dominant fission fuel for over six decades. However, uranium-235 makes up only about 0.7% of naturally occurring uranium. The remaining 99.3% is uranium-238, which is difficult to fission and gets in the way. To make an efficient bomb (i.e. one that makes the most of its fuel), we want as much fission to happen as quickly as possible. Remember, we are racing against the clock because the bomb rapidly blows itself apart. Therefore, it is important to maximize the number of neutrons from each fission that go on to cause a subsequent fission. This means that every U-238 atom in the fuel represents a trap that can steal away a neutron without causing fission. In fact, natural uranium has so much U-238 that it isn't fissile (i.e. it can't support a chain reaction). For this reason, we have to manually remove U-238 in a process known as *enrichment*. For weapons, we must enrich the uranium a lot, typically to be more than 85% U-235 (remember that we start at 0.7% U-235). As we will see, this turns out to be very difficult, leaving us wondering if there isn't an easier way.

Plutonium-239 and uranium-233 provide alternatives. In many ways, both are preferable to uranium-235. Figure 10.6 shows that they have somewhat larger fission cross-sections. Moreover, they produce more neutrons per fission, meaning that they can tolerate more of their neutrons getting lost. These considerations are quantified by the *critical mass* (see the Tech Box below), which you want to be as small as possible because it enables a weapon with a fixed size to have a larger yield.[13] Formally, the critical mass is the minimum amount

---

[13] For bomb design. For the survival of humanity, you would prefer the critical masses of all isotopes to be enormous.

of a given isotope required to sustain a fission chain reaction. If you don't have enough fuel, then so many neutrons will escape through the fuel's surface that the chain reaction will die. Uranium-235 has a critical mass of 52 kilograms, while uranium-233's is 15 kilograms and plutonium-239's is just 10 kilograms.[14] For this reason, plutonium-239 forms the majority of the world's nuclear weapons.

TECH BOX: Critical mass

Think about how the number of neutrons in a sphere of fuel would change with time, if a single fission event were triggered at its center.[15] On average, a plutonium-239 fission event releases almost 3 neutrons. So, if there is enough fuel located around the fission event, then at least one of these neutrons will go on to cause another fission. Hence, the chain reaction will be sustained. However, if we replace all the plutonium-239 with uranium-235, then things might change. The average uranium-235 fission only releases 2.4 neutrons and its fission cross-section is smaller. This means that you will need a bigger sphere of uranium-235 to sustain a chain reaction. The minimum mass of a sphere that still can sustain a chain reaction is termed the critical mass of the isotope.

There is one crucial problem with all three of our fuels: they are all very difficult to obtain. Uranium-235 is common in nature, but only appears distributed amongst a much larger amount of

---

[14]Californium-252 has a critical mass of just 2.7 kilograms! However, it is both one of the most dangerous and most expensive isotopes in existence. It's so keen to fission that, given a few years, it will do so spontaneously (no neutron needed). Just 1 gram of californium-252 will set you back over $10 million and produce *trillions* of neutrons per second (due to spontaneous fission and other forms of radioactive decay). It really is incredible how hostile to life this isotope is.

[15]A sphere is the right shape to use because it minimizes the surface area for a given volume. Hence, the number of neutrons leaving the material is minimized.

uranium-238. To make natural uranium usable, we must enrich it to artificially increase the fraction of uranium-235. Plutonium-239 and uranium-233, on the other hand, don't exist in nature. Instead, they must be bred from uranium-238 and thorium-232, respectively. Both of these options, enrichment and breeding, carry their own set of challenges that we will explore in the next two sections.

10.2.1 *Uranium enrichment*

How do we enrich uranium to increase the fraction of U-235? Since U-235 and U-238 are isotopes, they are chemically indistinguishable, so separating them using their chemical properties is impossible. Instead we must turn to other, much more challenging methods. While there are a number of theoretically possible techniques, only two have seen widespread application: gaseous diffusion and gas centrifuges. Both exploit the mass difference between U-235 and U-238. However, this difference is very slight, since three extra nucleons isn't much in a nucleus with 235.

Preparing for either gaseous diffusion or gas centrifuges requires a bit of work (see Figure 10.7). This is because raw uranium ore, which is what is dug out of the ground, contains as little as 0.1% uranium. The ore must be milled to concentrate the uranium. This produces *yellowcake*, which is over 80% uranium, but is a solid. The yellowcake is then converted into a gas called uranium hexafluoride, known as *hex* (which is shown in Figure 10.7). This form is appropriate for our enrichment techniques.

Gaseous diffusion works on the principle that two gases with different masses will pass through a small hole at different rates.[16] This means that a hex molecule with a U-235 atom will pass through 0.4% faster than a hex molecule with a U-238 atom. Therefore, you can start with hex on one side of a membrane with a bunch of holes and

---

[16]Formally, this is know as Graham's Law, which observes that

$$\frac{\text{flow rate of gas 1}}{\text{flow rate of gas 2}} = \sqrt{\frac{\text{molecular mass of gas 2}}{\text{molecular mass of gas 1}}}.$$

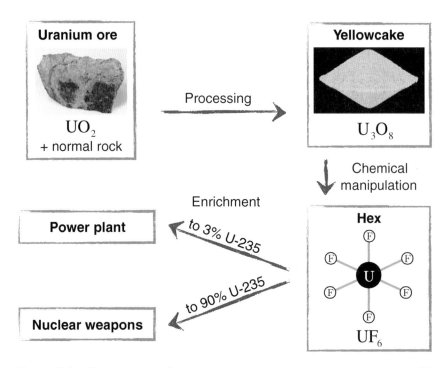

Figure 10.7: The process used to convert raw uranium ore into a material suitable for enrichment.

try to push it through. After a short time, you suck out the gas that has made it through the holes, which will contain a higher fraction of U-235. However, since the difference is so slight, this process must be repeated thousands of times to achieve a substantial degree of enrichment. Still, arduous as it is, this was the predominant method employed for enriching the uranium used in the Manhattan Project.

Currently, the overwhelming majority of the world's uranium is enriched using gas centrifuges. A centrifuge is a large, sealed cylinder that rotates about its center at a high rate.[17] When hex is injected in, the heavier molecules (i.e. those with a U-238 atom) are pushed outwards more strongly than the lighter molecules (i.e. those with

---

[17] For enrichment, centrifuges typically rotate roughly 500 times per second. The Stuxnet computer virus, which famously destroyed a significant fraction of Iran's centrifuges in 2010, subtly altered this rotational speed.

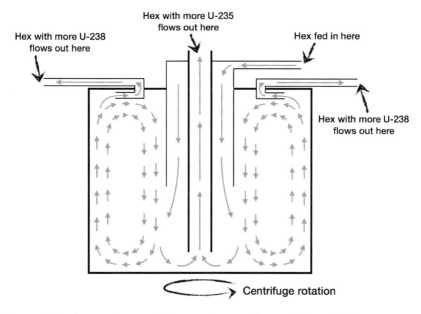

Figure 10.8: A centrifuge used for uranium enrichment. When the hex is inserted into the spinning chamber, the heavier gas molecules (i.e. those that contain a U-238 atom) preferentially move to the outside. This causes the concentration of U-235 in the hex near the center to be slightly larger, which can then be extracted and fed into another centrifuge for further enrichment.

a U-235 atom). The reason for this is the same as why spinning carnival rides require children to sit closer to the center than their parents. Otherwise, the parents (who are heavier and get pushed more strongly outwards) could squash their children. In a centrifuge, no squashing takes place, but the heavier molecules tend to force their way past lighter ones to end up closer to the outside of the device (see Figure 10.8). The gas that remains near the center (which has a slightly higher fraction of U-235) is then extracted and piped to the next centrifuge. As with gaseous diffusion, this process must be repeated many times to significantly increase the concentration of U-235.

Gas centrifuges have several advantages over gaseous diffusion. They are significantly more energy efficient, require less space, and can enrich material more quickly. The energy efficiency is particularly significant. Not only does this reduce cost, but it also makes enrichment plants harder to detect because they draw less electricity from the grid. This is important for a state that wants to clandestinely produce material for nuclear weapons.

In addition to centrifuges and gaseous diffusion, there is a third process called laser enrichment, which is approaching viability. It works on the principle that U-235 is slightly easier to ionize with a laser than U-238. Remember, ionization removes an electron from the atom, so it creates positively charged uranium atoms. Then, it is simple to turn on an electric field to attract the ionized uranium and separate it from the non-ionized uranium. Repeating this process many times will gradually increase the concentration of U-235. Laser enrichment is worrying because it is expected to require considerably less energy and space than diffusion or centrifugal enrichment, meaning it poses a larger proliferation concern.

However, for now we are fortunate. Enrichment is difficult, but is required because natural uranium has so little U-235. Enrichment is the primary obstacle preventing the construction of nuclear weapons. If the fractions of U-235 and U-238 in natural uranium were switched, modern civilization would likely be impossible. A small group of individuals could dig natural uranium out of the ground, build a rudimentary bomb, and detonate it in a city. Enrichment makes building nuclear weapons a country-sized endeavor, instead of an individual-sized one. Thank you Mother Nature!

10.2.2 *Plutonium production*

Unlike uranium-235, plutonium-239 decays away in a modest 24,000 years, so it does not occur naturally. Instead, it must be bred from U-238. This occurs via the reaction

$$^{238}\text{U} + {}^{1}\text{n} \rightarrow {}^{239}\text{U} \rightarrow {}^{239}\text{Np} \rightarrow {}^{239}\text{Pu}, \qquad (10.2)$$

where the uranium-238 captures a neutron, undergoes radioactive decay to neptunium-239, and then decays again to plutonium-239.[18]

---

[18]This occurs through a process known as "beta minus" decay. It takes a neutron and turns it into a proton and an electron. Also, note the awesome sequence of elements: *Ura*nium, *Neptu*nium, and *Pluto*nium. These names were concocted

However, it is unavoidable that, in the process of producing Pu-239, some Pu-240 will inadvertently be created. This is because, with neutrons flying around, some of the newly created Pu-239 will capture another neutron and transform into Pu-240. Since Pu-240 can spontaneously fission and release several neutrons, it is very risky for a nuclear weapon! The neutrons from spontaneous fission could set off a small chain reaction before the bomb properly detonates, leading to a "fizzle," rather than a "bang."[19] For this reason, so-called *weapons-grade* plutonium (i.e. plutonium that is over 90% Pu-239) must be made in specialized plutonium production reactors, which are designed to minimize the amount of Pu-240. This is done by putting uranium fuel in a neutron environment for short periods of time, usually around three months. If the fuel is left in for too long, then the plutonium will contain too much Pu-240. Still, we will see that plutonium production plants are fairly similar to conventional fission power plants.

Our final fissile fuel, uranium-233, is bred from thorium-232. However, it shares the same difficulties as plutonium. Specifically, its production inevitably creates other undesirable isotopes, such as uranium-232. Because U-233 has few advantages relative to Pu-239, but has the disadvantage of a 50% higher critical mass, it is rarely used in weapons.

10.2.3 *Weapon designs*

After obtaining the fissile material, making a crude nuclear bomb is relatively easy.[20] However, there are different levels of sophistication in nuclear weapon designs. But before we get to actual nuclear weapons, we must start with a simpler alternative: so-called *dirty bombs*. Strictly speaking, dirty bombs are not nuclear weapons because their explosion isn't driven by nuclear energy. Rather, they

---

before Pluto had been unmasked as a fraud. Had they been named today, who knows what plutonium would be called.
[19] Although the fizzle could still be quite big.
[20] We should mention that all of the information presented here is readily available on the internet. Even so, it would be impossible to build a nuclear weapon without years of research and a background in physics.

are conventional bombs that include radioactive material. Hence, dirty bombs have nowhere near the destructive power of nuclear weapons. Moreover, it is difficult to imagine a dirty bomb that causes widespread illness or loss of life from the radioactive material. The vast majority of the damage would be done by the conventional explosion. Instead, the primary motive for making a bomb dirty is to intensify public fear, complicate clean-up efforts, and raise the profile of the attack.

The most basic actual nuclear weapon is the gun-type bomb (shown in Figure 10.9). It consists of a "bullet" and a "target," both made of highly-enriched uranium. A chemical explosion is used to drive the bullet into the target. Once combined, they exceed the critical mass and explode due to energy released by the uncontrolled fission chain reaction. If a terrorist organization were to make a nuclear weapon, it would likely be of this design because of its simplicity and ease of manufacture. In a gun-type nuclear weapon, obtaining highly enriched uranium is the critical step.[21]

A second design is the implosion bomb. This setup is far more complex and expensive than the gun-type, but has a much higher yield.[22] There are several different configurations, but they all share a similar overall design. The device is composed of six concentric spheres. Layer six, the outermost layer, is the bomb casing needed to isolate the weapon from the outside world. Layer five is composed of chemical explosives that initiate the bomb. Their detonation creates inward-propagating shock waves that compresses the fuel (i.e. an implosion). Layer four is a "pusher," which is used to maximize the power of the inward-moving shock waves. Layer three is a U-238 "tamper," which slows down the expansion of the innermost layers after compression. This lengthens the time when fission is

---

[21]Because the bullet and target come together relatively slowly, it is exceedingly hard to make a gun-type plutonium weapon. The fission chain reaction in plutonium ramps up so quickly (due to spontaneous fission) that the device destroys itself before much fission can take place (i.e. a fizzle).

[22]Little Boy, the gun-type device dropped on Hiroshima, fissioned just over 1% of its fissile material. In contrast, Fat Man, the implosion-type device dropped on Nagasaki, fissioned around 13%. The implosion-type device is much more efficient because it can confine fissile material for longer and fission it faster.

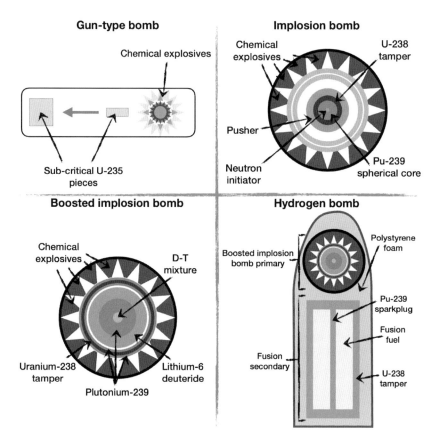

Figure 10.9: The designs of four types of nuclear weapons. A gun-type bomb (top left) fires one U-235 segment into another. An implosion bomb (top right) simultaneously detonates a large number of chemical explosives, collapsing the fissile fuel inwards. A boosted implosion bomb (bottom left) includes D–T fusion fuel in an implosion bomb to act as an additional source of neutrons. A hydrogen bomb (bottom right) uses the energy from a boosted implosion bomb to ignite a fusion burn.

possible, enhancing the yield of the weapon. The second layer is a spherical core of Pu-239 known as the "pit."[23] This is what fissions and produces most of the bomb's energy. The innermost layer, the

---

[23]Uranium-235 can be used here instead, but more is required since it has a higher critical mass than plutonium-239.

neutron initiator, is known as the "urchin." It is typically made of beryllium and provides a strong neutron source to kick-start the fission chain reaction.

Implosion bombs are also valuable in that they are the first step towards even more devastating devices: boosted and thermonuclear weapons. A boosted bomb has a similar design to an implosion-type device, but it adds several grams of *fusion* fuel. The fusion fuel, typically deuterium and/or tritium, produces large numbers of neutrons that boost the bomb's fission yield.[24] The fusion reactions contribute little to the overall energy released (usually less than a percent), but enhance the energy from fission. Moreover, boosting a weapon allows it to be made significantly smaller, a process known as miniaturization. This is problematic for proliferation because it enables the bombs to more easily fit onto ICBMs, which we know are difficult to defend against. However, it is exceedingly unlikely for terrorists to build boosted weapons. They simply don't have the technological expertise. Nowadays boosted weapons are ubiquitous, as they are used as a component in hydrogen bombs.

A *hydrogen bomb* (also known as a thermonuclear bomb) is designed to produce a large part of its energy output from fusion reactions. This contrasts with a boosted weapon, where the fusion fuel is used as a powerful neutron source to fission more of the fissile fuel. However, hydrogen bombs are much more difficult to construct. To ignite the fusion fuel, we must meet the Lawson criterion from Chapter 4. Physics doesn't give us a free pass just because it's a bomb. This means that, even at the optimal temperature, we still require the product of particle density and confinement time to exceed $n\tau_E \geq 2 \times 10^{20}$ seconds per cubic meter. Since it takes only about a microsecond (i.e. 10^{-6} seconds) for a bomb to explode, we see that the fuel must attain incredible plasma densities. It was recognized early on that only fission implosions, not chemical, could

---

[24] As of 2017, it has been reported that all U.S. nuclear armaments use tritium — a testament to its usefulness in nuclear weapons design. However, it should be noted that tritium is not necessary for a boosted nuclear weapon, only more effective at increasing yields.

achieve this extreme compression. To do this, a two-stage weapon, known as the Teller–Ulam design, was conceived. The basic idea was to first detonate a boosted implosion device, known as the primary. The electromagnetic radiation emitted from this would be used to compress and ignite the fusion secondary, releasing a vast amount of energy. The first ever hydrogen bomb, Ivy Mike, was detonated on November 1, 1952 and had a yield of 10 megatons — almost one thousand times more powerful than the gun-type bomb dropped on Hiroshima. However, despite their name, a large proportion of energy from hydrogen bombs stills comes from *fission*. Ivy Mike, for example, produced 77% of its yield from fission.

While pure fission weapons were able to level entire cities (see Figure 10.10), the amount of damage wrought by hydrogen weapons is unthinkable. A large thermonuclear bomb detonated over Boston, Massachusetts could break windows 100 kilometers away in Concord, New Hampshire. However, due to the nature of modern warfare, most nuclear weapons are miniaturized to be in the sub-megaton range. This allows as many as twelve nuclear warheads to be mounted on a single missile. Weapons with larger yields are more cumbersome, needing to be carried by bombers that can easily be intercepted before delivering their payloads.

Lastly, we note that all nuclear bombs require fissile material (see Figure 10.11). Currently, pure fusion weapons (i.e. nuclear weapons that do not require fissile material) appear far beyond our reach. In 2001, the US disclosed that it had "made a substantial investment in the past to develop a pure fusion weapon," but "does not have and is not developing a pure fusion weapon" and "no credible design for a pure fusion weapon resulted from the investment" [52].

10.3 Conventional Fission Reactors

Like nuclear bombs, generating electricity with fission power relies on a chain of fission reactions mediated by neutrons. By deliberately manipulating the number of neutrons, the total power of the chain reaction can be controlled. To keep the power constant, as is desirable in a power plant, exactly one neutron produced from each fission

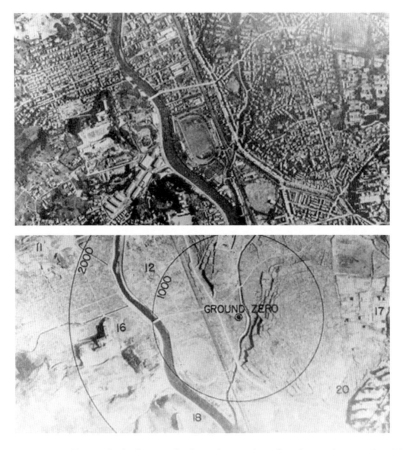

Figure 10.10: Nagasaki before and after the nuclear bomb on August 9, 1945. The distances labeling the circles in the bottom image are in feet.

event must go on to cause another fission event (on average). This is an important distinction from a bomb, where you want the power to spiral out of control. In a bomb, you would like *all* the neutrons produced by a fission reaction go on to cause more fission.

Because natural uranium is not fissile, a typical power plant *does* use enriched uranium fuel, but it only needs to be 3–5% U-235 (see Figure 10.12). It can make do with this low level of enrichment by using what is called a *moderator*. A moderator, usually normal water, causes the neutrons born from fission to scatter and slow down

Device	Materials required	Yield (approx.)
 Uranium gun-type bomb	• Uranium-235	20 kT
 Plutonium implosion bomb	• Plutonium-239 • Uranium-238	20 kT
 Boosted plutonium bomb	• Plutonium-239 • Uranium-238 • Lithium-6, Deuterium, and/or Tritium	< 1,000 kT
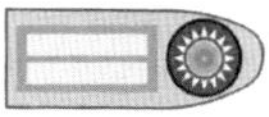 Hydrogen bomb	• Plutonium-239 or Uranium-235 • Uranium-238 • Lithium-6, Deuterium, and/or Tritium	50 kT to 50,000 kT

Figure 10.11: The fission and fusion materials required for different nuclear weapon designs.

without being absorbed.[25] Once the neutrons are slowed down, they can then take advantage of the much larger fission cross-section at

---

[25] The normal hydrogen in water does capture some neutrons, but not too many. A handful of reactors (notably Canadian CANDU reactors) chose to use D_2O (deuterated water) instead of normal water. Deuterium captures even fewer neutrons than normal hydrogen, which allows more neutrons to find uranium atoms to fission. This enables these reactors, called *heavy water reactors*, to run on natural uranium. The neutrons that deuterium does capture produce small quantities of tritium, which should be enough to startup the first fusion power plants.

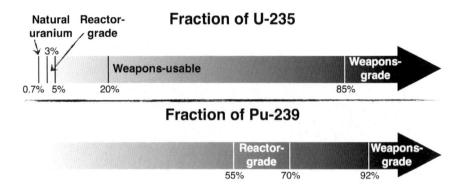

Figure 10.12: The level of enrichment needed for nuclear fission power generation versus weapons.

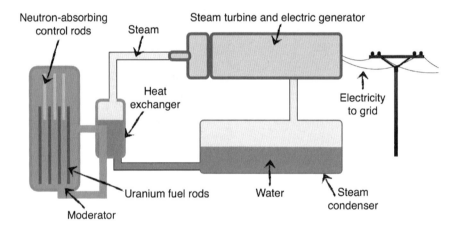

Figure 10.13: The basic components of a fission power plant.

low neutron energies (see Figure 10.6). This process also transfers the energy from the neutrons to the moderator, where it can be used to heat water, create steam, and drive a turbine to generate electricity (see Figure 10.13).

However, the enrichment needed to fuel a fission power plant is still a proliferation concern. This is because the technology is the same, regardless of whether you are enriching to 3% or 90%. In fact, making a bomb actually requires fewer centrifuges than fueling a fission power plant, even though it needs higher enrichment. This is because a bomb requires as little as 10 kilograms of enriched uranium, while operating a power plant for a year requires over 10,000 kilograms.

Another proliferation concern is posed by spent fuel, the high-level nuclear waste that is produced by fission reactors. This is what results after irradiating the U-235 and U-238 fuel with neutrons for a few years. When a neutron is absorbed by a uranium-235 or uranium-238 atom it can fission or sometimes it just keeps the neutron and transmutes into a heavier isotope like plutonium (see Figure 10.14). In fact, the production of plutonium can be responsible for up to 60% of the power output of a fission plant. What happens is the neutron environment created by the U-235 transmutes the U-238 into Pu-239, which then fissions and produces energy. Over half of all plutonium created in a reactor core is subsequently fissioned. However, we still must worry about what remains, which comes out as part of the spent fuel.

Fortunately, this *reactor-grade* plutonium (i.e. the plutonium that comes out of a fission power plant reactor) has a high fraction of

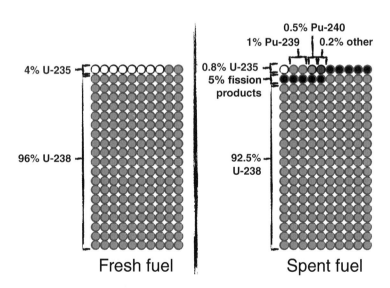

Figure 10.14: The composition of the low enrichment uranium fuel before (left) and after (right) it is put into a fission reactor.

Pu-240, which we learned makes it hard to directly use in a bomb. For this reason, power plants are not optimized to make weapons-grade material in the same way that a dedicated plutonium production plant is. That being said, simply taking the fuel out of a power plant reactor after a few months (instead of a few years) produces plutonium that is more relevant for weapons. In other words, it is important for nuclear security to monitor existing power plants and keep careful track of spent fuel.

10.4 Breeder Reactors

Standard fission reactors consume more fissile material than they generate. This results in them needing to be refueled periodically with fissile material. However, what if there were a reactor that could generate more fissile material than it consumes? Such reactors exist: breeder reactors. Beginning in the 1950s, when it was thought that uranium was fairly scarce, governments did substantial research into power plants that could produce fissile material on the fly. As more uranium reserves were discovered (and as uranium enrichment became increasingly affordable), interest in breeders declined.

A breeder reactor works by being fed *fertile* material, which is anything that can be transmuted into a fissile material by absorbing a neutron (e.g. U-238, Th-232). The neutron environment of the breeder then transmutes the fertile material into fissile fuel and fissions it. Conventional fission reactors already do this to some degree — they breed U-238 into Pu-239, which they fission. However, breeder power plants are optimized to increase this effect (see the following Tech Box).

TECH BOX: Breeder reactors

Neutrons from fission are born with high energies (about 1 MeV). Conventional reactors use a moderator to slow the neutrons down so that they are more likely to be absorbed by

---

(*Continued*)

(*Continued*)

fissile fuel like U-235 or Pu-239. This works because the fission cross-section is bigger at lower neutron energies. In a breeder, the moderator is usually reduced so that the neutrons take more time to slow down. This increases the probability that neutrons will cause breeding, rather than fission, because of relative sizes of the cross-sections at high neutron energy (see Figure 10.15). At these energies, the neutron capture cross-section of fertile atoms (e.g. U-238) is larger relative to the fission cross-section of fissile atoms. Hence, breeding reactions from neutron capture become more common. Note that the capture cross-section doesn't actually exceed the fission cross-section, but it doesn't need to. If there are many more fertile atoms than fissile atoms, they can overcome the fact that each individual fertile atom has a smaller cross-section.

Breeder reactors don't have to consume just uranium-238 or thorium-232. Since the fission that occurs inside them acts as a neutron source, they can process a whole range of different isotopes.

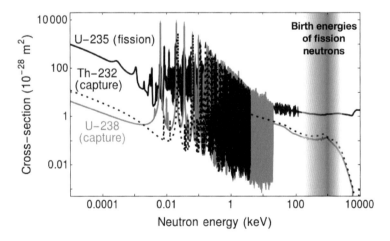

Figure 10.15: The neutron capture cross-section (i.e. the probability of a neutron being captured by a nucleus) for uranium-238 and thorium-232 compared to the fission cross-section of uranium-235.

This could be very useful, as they could burn through the spent fuel generated by conventional fission plants. Many advocates see this as recycling and hope that breeder reactors will replace conventional reactors once the technology becomes available. Unfortunately, as we will see in the next few sections, the nuclear security implications of fueling breeder reactors could be problematic.

TECH BOX: Fission–fusion hybrids

As a final note, there has been a lot of interesting work on energy generation using a combination of fission and fusion. The fundamental idea is that, instead of looking at a fusion reactor as a box that can generate electricity, we can view it as a box that produces a large number of neutrons. In conventional fusion systems, the idea is to use this neutron source to breed tritium from lithium (and produce a bit of extra energy). However, neutrons are useful for fission too. By surrounding a fusion reactor with fissile material, the neutrons will cause fission events and generate a bunch of energy (and more neutrons). However, because each fission event releases around 200 MeV and each fusion event only releases 17.6 MeV, it really still is a fission power plant. The advantage for fusion engineers is that their system doesn't have to produce net energy to be useful. The advantage for fission engineers is that, because the fission chain reaction is heavily dependent on the neutrons from fusion, they don't have to worry as much about the reaction spiraling out of control. The fusion part of the system is easy to shut off, so it can be used to kill the fission chain reaction. Other ideas for fission-fusion systems include using the fusion neutrons to breed fertile fuel into fissile fuel (like a breeder) or to burn the high-level radioactive waste produced by conventional fission plants. While these ideas could enable fusion systems to reach commercial viability sooner, their proliferation characteristics aren't much better than conventional fission power plants.

10.5 Fission Proliferation Risks

Continuing fission power generation as usual will increase the amount of spent nuclear fuel, consume our finite uranium supplies, and require continual flows of fissile material around the world. Breeder reactors can deal with the first two of these problems. They can actively consume the fissile material in nuclear waste and supply humanity's energy needs for millions of years (see Table 2.1 on page 59). Unfortunately, because of how they are fueled, their widespread construction poses an acute proliferation risk.

Currently, most nations simply take their spent nuclear fuel and set it aside.[26] Doing anything more is difficult. Spent fuel is called *self-protecting*, which means that it is so radioactive it can only be handled with radiation protection and sophisticated remote-handling equipment. Currently, the world has little capacity to do this. Breeders would change all of this. To fuel a breeder, we would need to be much more comfortable working with spent fuel. We would want to be able to separate it into its constituent elements and fabricate new fuel with a desired and well-characterized composition. So instead of spent fuel laying around in long-term storage too dangerous to approach, it would be handled, transported, manipulated, and fabricated. Once the fuel is made ready to fuel a breeder, it is no longer self-protecting and is largely weapons-ready. If someone were to steal it, they could use a *chemical* technique called PUREX to separate out the plutonium and/or enriched uranium needed to make a bomb. According to the International Atomic Energy Agency (IAEA), fuel for breeders would take roughly 1–3 weeks to be incorporated into a useable nuclear weapon. In contrast, the uranium used to fuel conventional fission reactors is still far from being weapon-ready — it requires further *enrichment*.

Moreover, while breeders reduce the amount of fissile material that must be stored long-term, they increase the total production of fissile material, roughly by a factor of three. This is primarily

---

[26]France, Japan, the United Kingdom, and Russia are notable exceptions.

because they generate so much more plutonium than conventional reactors. Additionally, since breeders produce minimal uranium, their widespread adoption would significantly alter the worldwide composition of fissile material.[27] This isn't necessarily good or bad, just different. The abundance of plutonium would make it easier for countries with weapons ambitions to build advanced bombs, but the reduction in enriched uranium would make it harder for nuclear terrorists to build low-tech gun-type weapons.

But before we get ahead of ourselves, what is the proliferation risk posed by the fissile material that we currently produce? To measure proliferation risk, we can use a unit called the *significant quantity* (SQ), defined as "the approximate amount of nuclear material for which the possibility of manufacturing a nuclear explosive device cannot be excluded."[28] So, if a terrorist got hold of one SQ worth of nuclear material they could make roughly one bomb. Currently, the world's reactors (which are almost entirely conventional) produce 20,000 SQs of spent fuel per year.[29] To put this in context, let's use the IAEA's uncertainty threshold, which is used for accounting material in spent fuel processing facilities. When the IAEA does inspections, the total amount of fissile material is required to be within 1% of expectations. So, even now, using conventional fission power plants to generate just 10% of the world's *electricity*, there is a worldwide margin of error of 200 potential bombs per year. But before you get too alarmed, remember that the amount of fissile

---

[27] Conventional reactors produce roughly equal parts Pu-239 and U-235.

[28] For uranium-235 this is roughly 25 kilograms, while for plutonium-239 and uranium-233 it is 8 kilograms. In contrast, the fuels for conventional reactors have much larger SQ values: low enrichment uranium is 75 kilograms and thorium is 20,000 kilograms.

[29] By volume, this isn't actually too much. For example, the spent fuel from 60 years of electricity generation in the US could fit in a single football field (if it was piled about 6 meters high). But the thing we really should care about is proliferation risk, not volume. Especially, given that, in the US, finding a centralized place to store spent fuel is a highly contentious and political issue. Instead, it is stored at the facility that generated it. From a security standpoint, it's better to ship it once to a centralized facility than leave it scattered around the country on some 100 odd sites.

material any single organization could conceivably skim off would be a small fraction of this.

Now imagine a hypothetical future in which we exclusively take the bravery option from Chapter 2 — we use breeder fission reactors to generate all of our *energy*. Since electricity is only about a third of all energy use and fission currently provides just 10% of that, satisfying our total energy needs would involve 30 times more fissile material than now. Moreover, by switching from conventional reactors to breeders, we roughly triple the production of fissile material. Hence, the worldwide margin of error becomes nearly 20,000 bombs *per year*, a factor of 100 more than our current situation. And much of this fissile material is not self-protecting.[30] Clearly, unless monitoring safeguards improve tremendously, nuclear power could have insurmountable proliferation problems.

Specifically, it is hard to see how arms reductions would be credible. Even with breeders and a modest expansion of nuclear power, the worlds largest energy producers could covertly amass nuclear fuel for a dozen nuclear weapons in just weeks. Over a few years, a country could conceivably construct hundreds, if not thousands of nuclear weapons. Since all countries would be aware of this fact, *a world without nuclear weapons, but with a large number of breeder fission reactors appears unrealistic.* There would always be the risk that one country would secretly skim a bit of fissile material off of their spent fuel processing facilities and create the world's only nuclear weapons.

In summary, we are presented with a difficult choice. Fission power plants could provide a sustainable low-carbon energy supply, but they endanger global nuclear security. We inevitably conclude that in the next few decades, we may have to make a difficult choice between building large numbers of fission reactors or continuing with fossil fuels. In other words, a choice between nuclear and climate security.

---

[30]If you are interested in specific details and more realistic scenarios, we recommend Ref. [53].

That being said, the dynamics of proliferation could change in the future, for better or worse. For example, laser implosion facilities like the National Ignition Facility (discussed in Chapter 9) can now imitate the detonation of nuclear weapons. From a weapons design perspective, they are of immense value. They reproduce the conditions in a bomb, without having to break international nuclear test ban treaties. If these bomb imitation facilities become increasingly ubiquitous, it may make it easier for countries to develop advanced nuclear weapons, particularly hydrogen bombs.[31] Another wildcard is the advent of laser enrichment facilities, which were mentioned earlier in this chapter. They look to have a far smaller footprint than centrifuges (both in terms of geography and energy), making them easier to hide. Generally speaking, any technology that allows fissile material enrichment faster and more covertly is bad for nuclear security.

On the other hand, technological advancement might improve the methods of detecting fissile material. This could enable the IAEA to reduce their margin of error and/or make enrichment and spent fuel processing facilities harder to conceal. Even something as simple as the ability to use cell phone cameras to detect the radiation signatures of fissile material could revolutionize nuclear security. If cell phone users were willing to give real-time access to the IAEA, it would become much harder to covertly move fissile material around the world.

Fundamentally, it is important that our ability to detect clandestine activity advances more quickly than our capacity to conceal it. Otherwise, the potential for nuclear attack will grow.

10.6 Fusion Proliferation Risks

Finally we arrive at the central question of this chapter: how would the advent of fusion reactors impact global nuclear security[32]?

---

[31]The absence of weapon imitation facilities would not curb the proliferation of simpler bombs. The gun-type bomb detonated over Hiroshima was done so without *any* prior testing — such was the Americans' confidence that it would work.

[32]For more details, we highly recommend Refs. [53–56].

Assuming they replace fission plants, positively! From a proliferation standpoint, there are several important characteristics of magnetic confinement fusion reactors using D–T fuel.[33]

The first and most significant characteristic: *there is no need for fissile or fertile material of any kind.* All isotopes of uranium, thorium, and plutonium can be entirely banned in fusion facilities worldwide. If fertile or fissile material was discovered, it could immediately be attributed to malicious intentions. No interpretation needed.

That said, since D–T fusion produces a large number of high energy neutrons, a fusion power plant could still be useful to someone seeking material for weapons. The neutrons from fusion can be used to breed fertile material into fissile material via the same reactions as in a fission breeder power plant (or a plutonium production plant). Practically, this could be achieved by injecting fertile material into the blanket surrounding the plasma, allowing it to capture the fusion neutrons. For example, inserting sub-millimeter spheres made of U-238 into liquid coolant flowing through the blanket would breed Pu-239. Fortunately, if we are smart about how we design, regulate, and monitor fusion power plants, this can be made very difficult.

Let's consider each of the three options to use a D–T fusion facility to obtain fissile material:

(1) produce material in a secret facility,
(2) secretly produce material in a declared facility, or
(3) openly produce material in a declared facility.

How would each of them play out?

The first option, concealing a fusion facility, seems nearly impossible. Like fission power plants, a fusion power plant would be conspicuous. They have a sizable geographic footprint (shown in Figure 10.16), require substantial amounts of electricity from the grid, and take ten or more years to construct. Spy satellites

---

[33] The distinction between different types of magnetic confinement fusion reactors is not very important for nuclear proliferation.

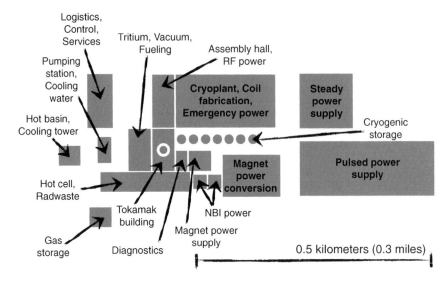

Figure 10.16: The general spatial layout of a fusion power facility like ITER. The specifics aren't important. The point is to give you a sense of how large the site is and the number of essential support facilities.

would notice them. Moreover, due to the multitude of essential support systems (e.g. cryogenics, fueling, cooling, power supplies, diagnostics), disabling a fusion power plant suspected of undergoing proliferation activities can be done safely. Indeed, it is arguable that fusion plants should be designed such that an external attack could be performed safely — an act of sincerity about the good intentions of the plant. Disabling the plant could be done without the release of radioactive material because an attack need not target the reactor. The most effective way would likely be to destroy the cryogenics plant needed to keep the magnets cool. And even if you were forced to destroy the reactor, the release of radioactive material from the fusion plant may well be tolerable.[34] This all is considerably different from a *fission* power plant, which is tricky to disable because it cannot be turned off quickly due to radioactive decay of the fission products.

---

[34]Obviously, releasing tritium (as well as activated material from the first wall) to the environment would still be horrible. This would be a last resort.

Fission plants must be disabled in such a way as not to interfere with its ability to safely shutdown. If you were to, say, bomb the cooling towers of a fission plant, it could inadvertently trigger a meltdown and large-scale release of radioactivity.

But what if someone tried to use a very small fusion system, rather than a power plant? The smallest conceptual design (which generates enough neutrons to pose a risk) is for a tokamak with a reactor diameter of around 1.6 meters and a fusion power of just 1.8 megawatts [57]. It isn't designed to produce net electricity. Instead, its purpose is to study the effects of D–T fusion neutrons. With optimistic performance, this reactor could generate enough neutrons to produce 1 SQ of U-233 or Pu-239 every 32 months. While this is concerning, all of the ancillary facilities (e.g. current drive, heating) require enough space and electricity from the grid to be detectable (around 40 MW). Additionally, preparing the fertile material for the fusion blanket and keeping it secret, requires a fairly extensive ground operation. Finally, the trace levels of tritium that would escape the reactor are detectable for tens of kilometers.[35]

Overall, it is extremely hard to hide even the smallest of magnetic confinement fusion devices. They require specialized components, technical expertise, electricity, and long construction times. So why bother? *A country with weapons aspirations would almost certainly find it much easier to covertly build an enrichment or plutonium production plant.*

How about the second option, secretly producing material in a declared fusion facility? In this case, we can assume that the fusion plant is being monitored by the IAEA.[36] The question then becomes if it is possible to produce a reasonable quantity of fissile material under the noses of inspectors. The most effective way to detect fissile material is to look for the gamma rays (i.e. very high energy

---

[35]When the Tokamak Fusion Test Reactor (TFTR) was running tritium experiments in Princeton, its operation could be detected by this method.

[36]Currently, all countries with nuclear facilities are members of the IAEA ... except for North Korea. But we don't need IAEA inspectors to know what's going on there.

photons) that it emits. Given the appropriate equipment, both U-238 and U-233 can be detected in minutes. This is so simple because there is no reason to have any fissile material whatsoever. In an enrichment facility or a *fission* power plant, there would necessarily be U-238 and/or U-233 all over the place, so detecting their gamma rays wouldn't be a red flag. We conclude that, as long as the fusion facility is being monitored, it is exceedingly difficult to produce fissile material covertly.

The third option is the most brazen — openly producing weapons-grade fissile material as quickly as possible using an existing fusion facility. Estimates indicate that a fusion power plant could be used to produce around 500 kilograms of plutonium-239 per year (which is similar to the amount generated in a breeder reactor). This means that it would take approximately six days to generate an SQ of plutonium. Even higher quantities of uranium-233 could be bred — as much as one SQ every two days! That's fast, especially if the proliferator takes steps to delay the detection of such material (which they would almost certainly do). Remember, the proliferator will know that their plant could be disabled rapidly.

Because of this scenario, any expulsion of inspectors from a fusion facility should be taken very seriously. There will need to be a well-defined framework to enable a *quick* response. However, it doesn't have to be a particularly sophisticated response. As soon as inspectors are expelled or sensors detect fissile material of any kind, the facility must be disabled.

While two days to make an SQ does seem short, it doesn't include the time needed for preparation. One of the key benefits of fusion relative to fission is that, at the start of illicit operations, not only is there no fissile material in the reactor, but the system isn't even set up to permit breeding. For example, it has been estimated that it would take no shorter than a month to modify a typical cooling system to allow small fertile particles to be introduced into the blanket [55]. And this doesn't include the time required to mine, process, and fabricate the fertile material off-site. In contrast, a fission reactor during normal operation already contains many SQs of material and has fertile material prepared. All the extra orchestration needed for breeding in a fusion device gives the world more margin for error. Finally, even after obtaining fissile material, it still must be fabricated into a bomb and delivered to its target. And by this point, the entire international community knows what is happening and is doing everything in its power to stop it.

While it is theoretically possible to use a fusion power plant to produce fissile material for use in nuclear weapons, it appears very difficult if common-sense precautions are taken. Fusion power certainly has better non-proliferation credentials than fission power. We also believe that the widespread use of fusion power can be made compatible with complete nuclear disarmament and a world entirely free of fissile material. To enable this, the world requires

(1) power plants that can be easily and safely disabled and
(2) an independent international regulatory agency with the ability to monitor for fertile material anywhere on site in real-time.

What do you think? Remember though, the above aren't required for the widespread adoption of fusion power. It's only necessary for the widespread adoption of fusion power *in a world without nuclear weapons*. In a world *with* nuclear weapons, why would bomb-makers bother with a fusion power plant? Countries will still have dedicated uranium enrichment and plutonium production facilities and there will still be fissile material to steal.

10.7 The Nuclear Energy Transition

After fusion power plants become available, it seems likely that there will be a period in which both fission and fusion power coexist. We will transition from an era of pure fission, through an era of fission and fusion, to an era of pure fusion.

As fusion power plants are built and fission plants are phased out, there will be a period of time when both fissile material and tritium will be abundant in the world (see Figure 10.17). Currently, there are only tens of kilograms of tritium in the world, so it is pretty hard to come by. But fusion power plants, each of which need around 100 kilograms just to start up, would change this. Most countries

	Fission materials			Fusion materials		
Energy mix	U-235	Pu-239	U-238	Lithium-6	Deuterium	Tritium
Pure conventional fission	Abundant	Abundant	Abundant	Abundant	Abundant	Scarce, but can be produced
Pure fission breeders	Potentially hard to obtain	Very abundant	Abundant	Abundant	Abundant	Scarce, but can be produced
Fission and fusion	Somewhat abundant	Somewhat abundant	Abundant	Abundant	Abundant	Abundant
Pure fusion	Potentially hard to obtain	Potentially hard to obtain	Abundant	Abundant	Abundant	Abundant

Figure 10.17: The availability of weapons-relevant materials for different ways of generating electricity. We are currently in the pure conventional fission row. Compare this information with Figure 10.11, which shows the material required for the different types of nuclear weapons.

will not have the capability to make this and will have to import it, leading to substantial amounts of tritium being shipped around the world. However, in practice, the increased availability of tritium is not as concerning as it might seem. Tritium is primarily useful in boosted implosion and hydrogen bombs, both of which require quite a bit of sophistication to build. Any organization that is capable of building these complex weapons would almost certainly be able to obtain tritium without much trouble. After all, it only takes a dozen *grams* of tritium to boost a fission bomb and it can already made in fission plants and particle accelerators (see the following Tech Box).

TECH BOX: Proliferating with tritium-boosted bombs

Let's say you are an evil country operating a fission reactor to make material for boosted implosion nuclear weapons.

(*Continued*)

> (*Continued*)
>
> How do you go about this? Well, a typical boosted weapon requires about 10 grams of tritium and 8,000 grams of plutonium. Additionally, a single atom of plutonium weighs about 80 times more than a single atom of tritium. Therefore, we see that a boosted bomb contains roughly 10 atoms of plutonium for each atom of tritium.
>
> Now, both plutonium and tritium can be bred in your fission reactor from a single neutron. If you decide to use one of your neutrons to make tritium, then you miss out on breeding a plutonium atom. However, since boosted weapons are 10 atoms of plutonium to 1 atom of tritium, you are devoting less than 10% of your neutrons to breeding tritium. So even if you could steal tritium from a fusion power plant (instead of having to breed it yourself), it would only increase your total bomb production by 10%. A 10% increase, while not ideal, does not seem like a massive problem.

Looking ahead, achieving a world at *global zero* (i.e. without any nuclear weapons) looks to be extremely difficult during both the fission and fission–fusion eras. In a world at global zero, the appearance of just a few nuclear warheads would confer enormous advantage to the aggressor. Therefore, a world without nuclear weapons, but with the fissile material needed for fission power, would be in a particularly fragile and high-stakes equilibrium. However, in a pure fusion era, it would be substantially harder to covertly build nuclear weapons because enrichment facilities and plutonium production plants could be entirely eliminated. Theoretically, fissile material could be banned worldwide. Admittedly, getting to a world with zero nuclear weapons under any conditions is devilishly hard. Fusion, by replacing the need for fission power, would make it easier.

10.8 Reshaping Geopolitics

The geopolitics of energy has become critical to national interests, especially in the past half century. Wars have been fought and governments destabilized in the pursuit of energy security. Fusion has the power to instill order here as well. Because, for fusion, the critical resources are intellectual and technological, rather than natural. This removes a major driver of global conflict. Developed countries that currently need to import large amounts of energy will save money and improve their energy security. Underdeveloped, fossil fuel-rich countries will escape from under the heels of more powerful countries. They can finally progress in peace, without the continual interference of third parties seeking to install a friendly regime.

Though harder to predict and quantify, these improvements in international relations would likely outstrip fusion's direct impact on nuclear security (i.e. reducing the amount of fissile material). After all, the reason that Sweden doesn't attack Norway isn't because it is too technically challenging to do covertly, it's because they have no desire to. An equitable and prosperous world will be a peaceful one, regardless of the accessibility of neutron sources.

10.9 Being a Role Model

In the modern era, scientific knowledge cannot be unlearned, nor is it easily constrained. When the Trinity test was detonated on July 16, 1945, civilization underwent an irreversible transition. On that day, civilization itself become a force of nature with the capacity for conscious self-destruction. Then and always, the world must have this burden on its mind.

However, nuclear energy has another side. A side with the capacity to provide humanity with large amounts of energy for billions of years. A side that can tackle climate change and, as we will soon see, potentially traverse interstellar distances. What fusion does is allow these two sides to be decoupled. With fusion, we can expand energy generation, while deliberately inhibiting our ability to construct nuclear weapons. We can select the best of both worlds. With sufficient willpower and global cooperation, we can entirely ban fissile material and leave novel enrichment techniques unexplored.

While achieving a supply of economic fusion power may finally bring about more robust nuclear and climate security, there is another benefit — being a good role model. A world without nuclear weapons, even though it has the knowledge to build them, is something to be admired. It would be a testament to human ingenuity, willpower, and vision. It demonstrates our ability to select only the best that science has to offer, an invaluable habit for the future. After all, the ability of technology to both empower and diminish civilization is not limited to nuclear physics.

Even if humanity escapes the nuclear age unscathed, the interlude will be temporary. Artificial intelligence, geo-engineering, and genetics are just a few examples of endeavors that pose great risks as well as great promise. Such is the price of progress. Our ability to handle the double-edged sword of nuclear energy will serve as a blueprint for the future. We must practice extracting the benefits of science, while suppressing its unsavory products. Undoubtedly, even sharper scientific and technological swords will come our way.

Chapter 11

Fusion and Space Exploration

The Earth is a precious lifeboat, floating in a dark and empty void utterly inhospitable to life. As of 2017, we have discovered thousands of planets apart from our own, but no signs of extraterrestrial life. Yet, even if the rest of the universe seems curiously devoid of life, some part of our species' future almost certainly lies out there. Curiosity, resources, and self-preservation are just a few of the many motivations to travel beyond our cradle known as Earth. Yet, this is not an easy task. Doing so requires traveling immense distances compared to anything from our everyday lives. At their closest, Mars is 55 million kilometers away, the Sun 150 million kilometers, Saturn 1.2 billion kilometers, and Pluto 4.3 billion kilometers. Unfortunately, short of science fiction (see Figure 11.1), physics doesn't allow us to get clever with our travel plans. This is probably already obvious to you, but the only way to travel immense distances within a human lifetime is to go really, really fast.

The distances in our Solar System may be large, but we have already sent probes to many of these locations. The fastest man-made object ever was NASA's Juno spacecraft, which used solar cells and chemical propulsion to accelerate to 165,000 miles per hour (relative to the Earth).[1] At this pace, we could travel to Pluto in

---

[1] To achieve this speed, Juno supplemented its on-board propulsion with a "gravity-assist" slingshot maneuver around the Earth.

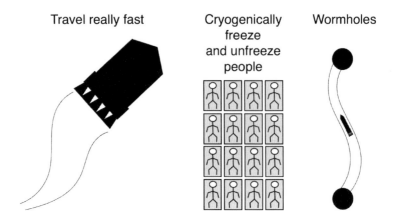

Figure 11.1: Some of our hypothetical options for interstellar travel. Currently, going fast looks to be the only one that will be practical anytime soon.

just a couple of years.[2] Thus, we see that our current propulsion techniques, sluggish as they are, nevertheless make it *possible* to navigate our Solar System.

The next step for humanity is a different story. The nearest star system to our own, Alpha Centauri, is 42 trillion kilometers away — 10,000 times further than Pluto. To cross this chasm, the Juno spacecraft would need tens of thousands of years. Nobody has time for that. On the other hand, if we could somehow travel very close to the speed of light, the ultimate speed limit of the universe, the trip could be done in as little as five years.[3] Clearly, interstellar travel requires spacecraft that can accelerate to a substantial fraction of the speed of light. While still very difficult, we will see that fusion may be able to do this. Just as the mastery of fire enabled cooking, a task that no amount of manual labor could accomplish, *fusion propulsion systems could enable humanity to do something that is currently impossible: the manned exploration other star systems.*

---

[2]We'd still recommend splurging for the emergency exit row seating. The extra leg room makes all the difference.

[3]Five years from the Earth's perspective — the travelers would actually experience a much shorter journey due to time dilation, a consequence of Einstein's special theory of relativity.

In addition to space travel, fusion would be particularly useful for space colonization. If we travel trillions of kilometers to another star system, there's no guarantee of what energy resources we would find there.[4] Fusion fuels contain the most energy per unit mass, so they would be the best to bring along. Additionally, fusion is very versatile as it can generate energy anytime, anywhere. But, most importantly, the fuels required for fusion are some of the most common elements in the universe. Hence, they are the energy resources that we would be most likely to stumble upon.

The focus of this chapter will be on space travel — how the advent of fusion power could open up the cosmos to human exploration.[5]

11.1 Basics of Spaceflight

How does one move in space? Unlike here on Earth, there is nothing to push against. Humans walk by pushing on the ground. A boat pushes water with its propeller. And an airplane pushes itself forward by using its jet engine to force air out the back of the plane. Since there is no air in outer space, a spacecraft has no such luxury. Instead, we must bring our own stuff along to push backwards out of the vehicle. This stuff is called *propellant* and it's a pain because we generally need a lot of it. To figure out how much, we must use *conservation of momentum*.

The momentum of an object is equal to its mass times its velocity — the faster something is going or the heavier it is, the larger its momentum. Now, imagine a spacecraft carrying a bunch of propellant sitting at rest somewhere in Outer Space. The spacecraft and the propellant both have zero momentum and conservation of momentum says that the total momentum must remain unchanged. Thus, if we force the propellant out of the vehicle in a particular

---

[4] Strictly speaking, by the definition of "traveling to a *star* system," there would be solar energy. However, problems could still arise if the planet we hoped to colonize turned out to have weather similar to the UK. Imagine that! Might as well turn around and start heading back to Earth.

[5] To learn more about space travel (and fusion's place in it), we recommend *Entering Space* by Robert Zubrin.

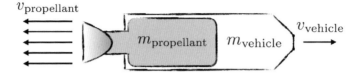

Figure 11.2: A simple schematic of a spacecraft.

direction, part of the total mass will have left, moving at some speed. Since the total momentum can't change, the vehicle must accelerate in the opposite direction (see Figure 11.2). Another equivalent way of looking at this is that the spacecraft has pushed off of the propellant in order to accelerate itself forwards. A common analogy is that of being stranded on the frictionless surface a frozen lake. You cannot get up because the lack of friction will cause you to immediately fall. How do you get off the lake? You take a shoe, or other heavy object, and throw it as hard as you can in the opposite direction to which you want to go. This will make you glide to the shore. Similarly, to make a spacecraft accelerate forwards rapidly, we want to push a large mass of propellant at a very high velocity out the back.

There is, however, a subtlety that makes the propellant velocity more important than the mass of propellant. If we decide to settle for a low propellant velocity and try to compensate with a very large mass of propellant, we will get frustrated. This is because, when we first start accelerating our spacecraft, we also have to accelerate all of the propellant that the vehicle is carrying. This means that simply loading up our spacecraft with more and more propellant produces diminishing returns, especially when the total mass of propellant starts to substantially exceed the mass of the vehicle (see the following Tech Box). You end up spending most of your propellant accelerating the rest of your propellant!

The propellant exhaust velocity is why fusion has the potential to be so useful. Conventional spacecraft, like those that took humans to the Moon, use chemical combustion for propulsion. Because of the low energy of chemical reactions, they can only eject propellant at around 0.001% the speed of light. In contrast, fusion reactions produce high energy particles that can be directly exhausted at 5%

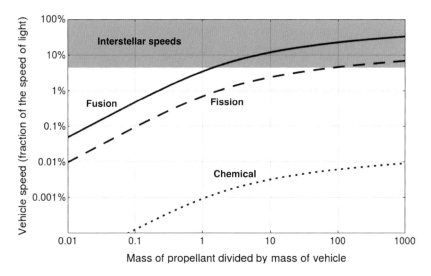

Figure 11.3: The amount of propellant needed to accelerate a spacecraft to a particular speed for the propellant exhaust velocities expected of fusion (5% of the speed of light), fission (1% of the speed of light), and chemical propulsion (0.001% of the speed of light).

the speed of light. The enormous impact of this is illustrated in Figure 11.3.

TECH BOX: The rocket equation

Working through the mathematical details of momentum conservation produces the *rocket equation*[6]:

$$m_{\text{propellant}} = m_{\text{vehicle}} \left(e^{v_{\text{vehicle}}/v_{\text{propellant}}} - 1 \right). \qquad (11.1)$$

Using this equation, we can calculate the mass of propellant $m_{\text{propellant}}$ needed to increase the velocity of a vehicle with a mass m_{vehicle} by an amount v_{vehicle}. Here $v_{\text{propellant}}$ is the

---

(*Continued*)

---

[6]This equation would need to be modified if external forces were present (e.g. "gravity-assist" slingshot maneuvers using planets) as well as if the velocities approached the speed of light (due to special relativity).

(*Continued*)

> exhaust velocity of the propellant and $e \approx 2.7183$ is a mathematical constant. We see that, if we want to accelerate the spacecraft from rest to move as fast as its propellant, then it must carry propellant weighing 1.7183 times more than the vehicle itself. This demonstrates the crucial importance of exhaust velocity. The faster it is, the less propellant you need to accelerate to a given speed — and the dependence is exponential!

Looking at Figure 11.3, you can see that a good rule-of-thumb is *the highest velocity that a spacecraft can reasonably attain is about twice its propellant velocity*. Going much faster requires truly absurd amounts of propellant. Hence, chemical propulsion, with a propellant velocity of 0.001%, is utterly unworkable for interstellar travel. Fission power, while dramatically better, still falls a bit short. Even with very optimistic assumptions about its exhaust velocity, a spacecraft that was 90% propellant would only reach 2% of the speed of light.[7] Factoring in the time needed for acceleration/deceleration, it would take over 200 years to reach Alpha Centauri. In comparison, the same spaceship using fusion could attain 10% of the speed of light and make the journey in roughly 75 years.[8] Still a *very* long time, but it's close to the limit of what could be workable (e.g. no one born on the spacecraft would be certain to die on it).

Another, surprising consideration is the sheer magnitude of energy required for a single interstellar journey. In Figure 11.4, we

---

[7] We say optimistic because it appears difficult to directly exhaust the high energy particles produced by fission. The neutrons are neutral, so it is difficult to collimate them effectively. Additionally, the stopping distance of fission products in solid material is much shorter than for neutrons, so exhausting the fission products while still sustaining a fission chain reaction looks impossible.

[8] Theoretically, matter–antimatter annihilation could complete the journey in 10 years or so, but antimatter propulsion is still very much science fiction. Currently, antimatter cannot be produced nor stored in any practical quantity.

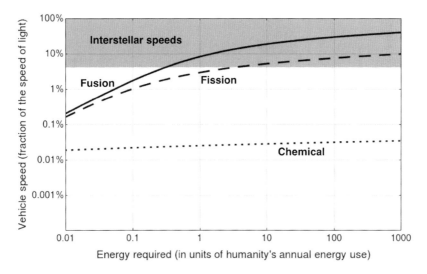

Figure 11.4: The energy required to accelerate a small spacecraft with the mass of a Boeing 747 airplane (i.e. a measly 400 tonnes) to a particular speed, given the propellant exhaust velocities from Figure 11.3.

see that accelerating a small spacecraft to 10% of the speed of light using fusion requires as much energy as humanity currently uses in a year. While this is a tall order, the world has more than enough fusion fuel. The primary challenge is to design a light-weight fusion thruster that can generate this amount of power. To burn though humanity's annual energy use over the course of 75 years would require the fusion thruster to operate at 250 GW. This power output is equivalent to 500 copies of ITER. Clearly, this level of performance is very ambitious (and still quite speculative) by today's standards. We require enormous power to be produced in light-weight devices (although they can have enormous volumes). Nevertheless, not only is fusion our best option for manned interstellar travel, it currently appears to be our only one.[9]

---

[9] There are other ways of getting exceedingly light spacecraft to interstellar speeds. For example, Breakthrough Starshot is a project currently in development that aims to accelerate tiny probes to around 20% the speed of light by pushing them with powerful Earth-based lasers. However, these probes are envisioned to weigh just a few grams.

11.2 Fusion Thruster

As discussed, designing an interstellar fusion thruster makes a commercial fusion power plant look easy. We have just seen that, even in optimistic circumstances, travel to Alpha Centauri would take an uncomfortably long time. This means that a fusion thruster has little room for inefficiency. Moreover, throughout this book, we have focused on deuterium–tritium fusion because this reaction is significantly easier than anything else. Unfortunately, D–T fusion is not typically used in conceptual spacecraft designs. This is primarily because 80% of the energy produced is carried away by neutrons. Not only does this necessitate thick, massive shielding, but neutrons cannot be effectively redirected out of the back of the spacecraft to maximize thrust. Additionally, since tritium is radioactive, half of a spacecraft's fuel stockpile will vanish every 12 years. The alternative would be to bring lithium and breed tritium along the way, but lithium is twice as heavy.

For these reasons, interstellar propulsion systems are generally imagined to use either D–D or D–^{3}He fusion. Both reactions produce most of their energy in the form of charged particles, which can be manipulated into beams with magnetic fields to maximize thrust. The ideal reaction is the helium-3 one because each fusion produces a whopping 18.3 MeV of energy, entirely carried by charged particles. Thus, it needs minimal shielding and all of the energy can be directly used to produce thrust. Theoretically, D–^{3}He could achieve propellant velocities of 9% of the speed of light. The difficulty is that helium-3 is not abundant here on Earth, though it is available elsewhere in our Solar System (e.g. the Moon, Jupiter). The practicality of accessing these extraterrestrial resources, however, is unclear.

The alternative, D–D fusion, should cause no concerns about fuel shortages — in Chapter 2 we saw that it was the dominant stockpile of energy on Earth. Unfortunately, it has an even lower fusion cross-section than D–^{3}He. Interestingly, the D–D reaction has two branches,

$$2\,^2\text{H} \begin{matrix} \nearrow\,^3\text{H} + \,^1\text{H} + 4.0\,\text{MeV} \\ \searrow\,^3\text{He} + \,^1\text{n} + 3.3\,\text{MeV}, \end{matrix} \tag{11.2}$$

which produce both tritium and helium-3. Thus, if we have a device capable of D–D fusion, then we can also use it to fuse the products in D–T and D–^{3}He reactions. This is called a *catalyzed* D–D fuel cycle and it produces 60% of its energy in the form of charged particles. While not quite as good as D–^{3}He, a catalyzed D–D fuel cycle still has a theoretical propellant velocity of 7% of the speed of light (and all the fuel is found on Earth).

To implement these fuel cycles, a large number of different thruster designs have been proposed. However, considering that we don't yet have a D–T device that works on Earth, committing to a spacecraft design is entirely premature. Nevertheless, for illustrative purposes, we show one of the most elegant concepts in Figure 11.5. This design is built around the magnetic mirror that we learned about way back in Chapter 4. If you remember, magnetic mirrors are linear devices that have two ends, which are pinched closed by strong magnetic fields, but tend to leak particles. If we could overcome their confinement difficulties, magnetic mirrors would make for quite nice thrusters. We could achieve fusion in the middle of the device and, by independently adjusting the magnetic field strength at the two ends, control the leak rate. Particles exiting one end of the device could be used for propulsion, while particles exiting the other could generate electricity for the spacecraft. Moreover, because of the way the magnetic mirror force works, the particles that leak out are the ones that have velocities almost entirely parallel to the magnetic

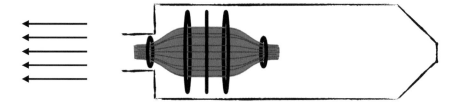

field. Hence, the fusion products will exit the device already focused into a beam — ideal for maximizing thrust.

While fusion clearly has a very long way to go before it will be powering an interstellar colonization mission, it is the only technology that appears capable of such a feat. Moreover, interstellar travel is an extremely high bar that fusion thrusters don't need to meet to be useful. Even a lightweight D–T tokamak, by virtue of its energy density, has the potential make travel within our Solar System much easier. It could be used to explore, mine resources, better defend against asteroids, or even to colonize Mars. Nevertheless, for some reason, interstellar travel seems like the goal to aim for. After all, a wise man once said "Shoot for the moon. Even if you miss, you'll land among the stars." Wait a second ... no, that's not right at all.

PART 5
CONCLUSIONS

Chapter 12

When Will We Have Fusion?

In the next 100 years or so, humanity will be forced into an energy revolution. Whether it be due to climate change or fuel shortages, *there will be immense change.* Currently, 80% of our energy supply is provided by fossil fuels and must be entirely replaced. Unfortunately, we have seen that our options are quite limited. There are only three energy sources that hold the potential for sustainability: renewables, nuclear fission, and/or fusion. All three look challenging. Renewables require cheap energy storage and millions of enormous man-made structures covering much of the globe. Nuclear fission requires a transformation in public relations and a tolerance for the risks posed by nuclear proliferation. And fusion, while ideal, is really difficult to get working. Thus, we are not in a position to be picky. These three are not options for us to choose among, but strategies to be aggressively pursued.

Since fusion research began in the 1950s, there has been enormous progress. For several decades, the performance of magnetic confinement fusion devices improved more rapidly than computing technology. Breakthroughs like H-mode, plasma shaping, and superconducting magnets have made an economically viable power plant much more achievable. Moreover, since fusion is such a messy problem that spans so many disparate areas of science and technology, it seems almost certain that human ingenuity will eventually prevail. What's more important, however, is "when?"

The response by critics is often a joke: "fusion is 20 years away . . . and always will be." The true answer is even less funny: we'll have fusion soon after people start taking it seriously. This is because progress in fusion depends more on society than it does on fusion researchers. The proximity to economically viable fusion power should not be measured in years — it should be measured in dollars. To this end, in 1976, the US Energy Research and Development Administration published a magnetic confinement fusion program plan [58]. In it, they laid out four different paths the nation could take to produce the first demonstration power plant (see Figure 12.1). While some paths took longer than others and varied in terms of their *yearly* spending, each path roughly cost $100 billion in total. This report also identified a fifth plan, which had such low funding that "a practical fusion power system might never be built." This "maybe never" plan had an annual budget five times smaller than the others, enough to maintain the workforce and little more. As you can probably guess, the actual level of spending has been far short of even the "maybe never" plan.

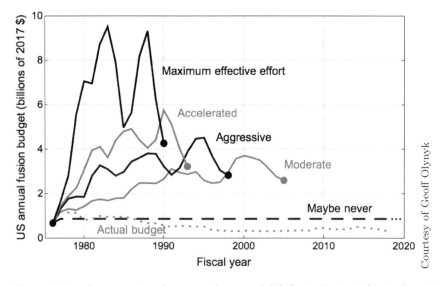

Figure 12.1: A comparison between the actual US fusion budget (dotted) and four different funding plans predicted to produce a demonstration power plant on different timescales (solid). Also shown is a funding scheme that may never lead to a power plant (dashed).

The consequences of this are no better illustrated than by our own personal experiences. America was where one of us was born and where both of us studied as undergraduates. It was there that we became interested in fusion, but sadly neither of us could remain for long. In 2012, faced with 10 or 15 years of substantial contributions to ITER, Congress decided to shutdown Alcator C-Mod, one of the three major national tokamaks. This decision, a consequence of dwindling domestic fusion budgets, directly resulted in both of us relocating and many of our colleagues leaving the field. This was tragic and symptomatic of a broader trend.

Since the 1980s, the epicenter of fusion research has been gradually moving east. The siting of ITER in France embodies the near-term dominance of the European program. Despite setbacks, ITER is steadily progressing toward its goal of demonstrating the scientific and technological feasibility of a tokamak fusion power plant. It's really hard to overstate its importance. The success of ITER would be a milestone of human civilization and could trigger a worldwide race to a demonstration power plant. On the other hand, if ITER fails (or fails to be completed), it seems certain to delay fusion by decades. For at least the next 15 years, ITER is everything for the community.

Looking forward, it is clear that the countries with the biggest fusion ambitions are in East Asia. South Korea already has a date in mind for its own power plant, K-DEMO. China is planning a new domestic tokamak that will be even larger than ITER. Moreover, JT-60U in Japan already holds the record for best plasma performance and will soon be replaced with the even more capable JT-60SA. It would be good for the world if every country could keep up. Throughout human history, advances in energy technology have been intrinsically linked with a general improvement in the human condition. Not only would fusion massively increase the amount of energy available, but it would remove strong drives of global conflict.

To this end, a small shift in public support, anywhere in the democratic world, would significantly improve the prospects for fusion. Though some places are better than others, spending on fusion is fairly minimal everywhere. Since 1976, the entire world

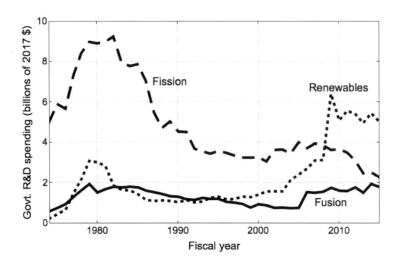

Figure 12.2: The estimated worldwide governmental spending on the research and development of fusion, fission, and renewables (including energy storage and transmission)[1] [59]. Research and development financed by private companies looks to roughly double fission and renewables spending, but is negligible for fusion [60, 61].

has spent about $60 billion on fusion research, roughly $1.5 billion per year. This is barely half of what the 1976 report estimated would produce a power plant and dramatically less than fission and renewables (see Figure 12.2). Through taxes, the average citizen in the developed world contributes about $1.50 per year to fusion research. Increasing this to $2 would allow the world to properly fund ITER without making cuts to domestic programs and would only cost the world as much as buying a few jumbo jets each year. An increase to $7 would make fusion spending comparable to fission and renewables and put the world on the most ambitious path in Figure 12.1. Again, because funding for fusion is already so low, accomplishing this doesn't require a major shift in political priorities, just a small group of vocal advocates. So don't underestimate your power to influence the priorities of elected officials. If you wish,

---

[1] This data is only for the countries that belong to the IEA, which includes most countries with substantial fusion programs (except for China, Russia, and India).

contact your representatives and tell them what you think. Tell them that fusion would make us energy independent or safer from nuclear attacks or better stewards of our planet. Tell them why you think it's important to fund fusion research and development. It may very well be the first they've heard of the program.

Finally, if you are a young person still exploring careers,[2] we hope that we have instilled in you our fascination with the challenges of fusion. As you have seen, there are many outstanding problems that require many different skills, all of which need your attention. Fusion is hard, but it is not so hard that it won't yield to ingenuity and diligence. Humanity has harnessed the power of fire and animals, wind and water, and sunlight and fission. *Fusion is next.* Regardless of whether you're a scientist, civil servant, businessperson, engineer, or simply someone interested in fusion, we hope that you will do whatever you can to help with one of the most exciting and consequential endeavors in human history.

---

[2] Or an older person still exploring careers.

American-style donuts

Prep time	**Cook time**	**Serves**
∼ 20 years	100 million years	7.6 billion people

- 4 cups flour
- 1 teaspoon salt
- 8 fluid ounces milk
- 5 cubic inches sugar
- 1/5000 bushel yeast
- 1/2 gill butter
- 2 eggs
- vegetable oil
- granulated sugar

Instructions
(1) Mix the flour, salt, milk, sugar, yeast, butter, and eggs and knead until homogeneous.
(2) Let rise in an oiled bowl for one hour or until too large to be managed by any single country.
(3) Punch down in size until the US rejoins the project, then let rise for another hour.
(4) Roll out dough on a floured surface into a layer about 1/2 inch thick.
(5) Cut the dough into toroids and leave to rise for half an hour.
(6) Fry in pure vegetable oil (taking care to minimize impurities) at 200,000,000°F for 3 minutes.
(7) Let cool (three energy confinement times at a minimum) and roll in granulated sugar.

French crullers

Photo by Adrienne Zwart Photography

Prep time
> 20 years

Cook time
10 billion years

Serves
7.6 billion people

Ingredients
- 125 grams flour
- 1 gram salt
- 250 grams heavy water (do NOT substitute with tritiated water)*
- 10 grams sugar
- 75 grams butter
- 175 grams egg
- 50 grams egg whites
- vegetable oil (in grams)

Instructions
(1) Bring the salt, water, sugar, and butter to a boil in a saucepan.
(2) Stir in the flour and keep stirring until the mixture starts to coat the pan.
(3) Move mixture to a bowl and slowly stir in the eggs and egg whites until it begins to hold a shape.
(4) Form the dough into the optimal shape to retain heat (how to do this in practice is left as an exercise for the chef).
(5) Fry in vegetable oil at 100,000,000°C for 4 minutes or until net energy production is achieved.

* But seriously, heavy water can really mess you up too, so don't use it either.

Powdered sugar donut holes [UNCLASSIFIED]

Prep time	**Cook time**	**Serves**
Allegedly finished	1 nanosecond	7.6 billion people

Ingredients
- 4 cups flour
- 1 teaspoon salt
- 8 fluid ounces milk
- 5 cubic inches sugar
- 1/5000 bushel yeast
- 1/2 gill butter
- 2 eggs
- vegetable oil
- Powdered sugar

Instructions
(1) Follow steps (1) through (4) of the above donut recipe.
(2) Shape the dough into small balls, ensuring that each is smooth to 1 part in a 10,000 (otherwise they will not cook evenly).
(3) Cook using at least 2 MJ of energy, but almost certainly less than 100 MJ (for more consistent results cook using a nuclear bomb, but this may affect flavor).
(4) Let cool to scientific breakeven (if reached) and roll in powdered sugar.

Bibliography

[1] T. Bruckner, I. Bashmakov *et al.*, Chapter 7 — Energy systems, Tech. rep., IPCC (2014).
[2] A. Bradshaw, T. Hamacher & U. Fischer, Is nuclear fusion a sustainable energy form? *Fusion Engineering and Design* **86**, 9–11, pp. 2770–2773 (2011).
[3] R. Chester, *Marine Geochemistry*. Blackwell Science (2000).
[4] M. Tamada, N. Seko *et al.*, Cost estimation of uranium recovery from seawater with system of braid type adsorbent, *Transactions of the Atomic Energy Society of Japan* **5**, 4, pp. 358–363 (2006).
[5] World Energy Council, 2007 Survey of energy resources, *Survey of Energy Resources* p. 381 (2007).
[6] Lazard, Lazard's levelized cost of energy analysis 10.0, *Lazard's LCOE*, December 2016, pp. 1–21 (2016).
[7] W. Moomaw, P. Burgherr *et al.*, Annex II: Methodology, *IPCC Special Report on Renewable Energy Sources and Climate Change Mitigation* (2014).
[8] A. De Vos & P. van der Wel, The efficiency of the conversion of solar energy into wind energy by means of Hadley cells, *Theoretical and Applied Climatology* **46**, 4, pp. 193–202 (1993).
[9] A. S. Adams & D. W. Keith, Are global wind power resource estimates overstated? *Environmental Research Letters* **8**, 1 (2013).
[10] Netztransparenz, Windenergie Hochrechnung, (2017), netztransparenz.de.
[11] European Network of Transmission System Operators for Electricity, Monthly Hourly Load, (2016).
[12] F. Wagner, Surplus from and storage of electricity generated by intermittent sources, *The European Physical Journal Plus* **131**, 12, p. 445 (2016).
[13] A. O. Converse, Seasonal energy storage in a renewable energy system, *Proceedings of the IEEE* **100**, 2, pp. 401–409 (2012).
[14] B. Vest, Levelized cost and levelized avoided cost of new generation resources in the annual energy outlook 2016, *US EIA LCOE*, August, pp. 1–20 (2016).
[15] P. F. Bach, 2016 wind energy data, (2016), http://www.pfbach.dk/.
[16] D. J. C. MacKay, *Sustainable Energy — Without the Hot Air*. UIT Cambridge, Cambridge (2009).

[17] IAEA, ENDF Database, (2017).
[18] T. H. Stix, Heating of toroidal plasmas by neutral injection, *Plasma Physics* **14**, 4, p. 367 (1972).
[19] J. D. Lawson, Some criteria for a power producing thermonuclear reactor, *Proceedings of the Physical Society Section B* **70**, 1, pp. 6–10 (1957).
[20] M. Keilhacker, A. Gibson et al., High fusion performance from deuterium-tritium plasmas in JET, *Nuclear Fusion* **39**, 2, pp. 209–234 (1999).
[21] A. J. Webster, Fusion: Power for the future, *Physics Education* **38**, 2, pp. 135–142 (2003).
[22] T. Sunn Pedersen, A. Dinklage et al., Key results from the first plasma operation phase and outlook for future performance in Wendelstein 7-X, *Physics of Plasmas* **24**, 5, p. 055503 (2017).
[23] V. Riccardo, S. Walker & P. Noll, Modelling magnetic forces during asymmetric vertical displacement events in JET, *Fusion Engineering and Design* **47**, 4, pp. 389–402 (2000).
[24] H. A. Bethe, Energy production in stars, *Physical Review* **55**, 5, p. 434 (1939).
[25] F. Wagner, G. Becker et al., Regime of improved confinement and high beta in neutral-beam-heated divertor discharges of the ASDEX tokamak, *Physical Review Letters* **49**, 19, p. 1408 (1982).
[26] P. B. Snyder, H. R. Wilson et al., Edge localized modes and the pedestal: A model based on coupled peeling–ballooning modes, *Physics of Plasmas* **9**, 5, pp. 2037–2043 (2002).
[27] P. Snyder, R. Groebner et al., A first-principles predictive model of the pedestal height and width: Development, testing and ITER optimization with the EPED model, *Nuclear Fusion* **51**, 10, p. 103016 (2011).
[28] J. Jacquinot et al., Deuterium-tritium operation in magnetic confinement experiments: Results and underlying physics, *Plasma Physics and Controlled Fusion* **41**, p. A13 (1999).
[29] X. Litaudon, JET program for closing gaps to fusion energy, *IEEE Transactions on Plasma Science* **44**, 9, pp. 1481–1488 (2016).
[30] G. G. Howes, S. C. Cowley et al., Astrophysical gyrokinetics: Basic equations and linear theory, *The Astrophysical Journal* **651**, 1, pp. 590–614 (2006).
[31] B. Sorbom, J. Ball et al., ARC: A compact, high-field, fusion nuclear science facility and demonstration power plant with demountable magnets, *Fusion Engineering and Design* **100**, 378 (2015).
[32] W. Broad, Secret advance in nuclear fusion spurs a dispute among scientists, (21 March 1988).
[33] A. Schaper, Arms control at the stage of research and development? — the case of inertial confinement fusion, (1991).
[34] R. Steinhaus & M. Moynihan, Full interview — Mr. Robert Steinhaus, (2017), www.thefusionpodcast.com/full-interviews/mr-robert-steinhaus, podcast.
[35] L. Berzak Hopkins, S. LePape et al., Design and follow-on from ~50 kJ fusion yield using high-density carbon capsules at the National Ignition Facility, in

59th Annual Meeting of the American Physical Society Division of Plasma Physics, 12 (2017).

[36] M. Kuriyama, N. Akino *et al.*, Operation and development of the 500-keV negative-ion-based neutral beam injection system for JT-60U, *Fusion Science and Technology* **42**, 2–3, pp. 410–423 (2002).

[37] C. Ebbers & E. Moses, The mercury laser system — A scaleable average-power laser for fusion and beyond, Tech. Rep. LLNL-JRNL-402612, Lawrence Livermore National Laboratory (2008).

[38] O. Hurricane, D. Callahan *et al.*, Fuel gain exceeding unity in an inertially confined fusion implosion, *Nature* **506**, 7488, p. 343 (2014).

[39] J. Slough, G. Votroubek & C. Pihl, Creation of a high-temperature plasma through merging and compression of supersonic field reversed configuration plasmoids, *Nuclear Fusion* **51**, 5 (2011).

[40] T. J. McGuire, The Lockheed Martin compact fusion reactor, (2015), oral presentation at Princeton University.

[41] H. Gota, M. Binderbauer *et al.*, Achievement of field-reversed configuration plasma sustainment via 10 MW neutral-beam injection on the C-2U device, *Nuclear Fusion* **57**, 11, p. 116021 (2017).

[42] S. Howard, Experimental results from the SPECTOR device at General Fusion, (2017), oral presentation at 27th IEEE Symposium on Fusion Engineering.

[43] T. J. McGuire, G. Font *et al.*, Lockheed Martin compact fusion reactor concept, confinement model and T4B experiment, in *58th Annual Meeting of the American Physical Society Division of Plasma Physics*, 18 (2016).

[44] T. J. McGuire, Encapsulating magnetic fields for plasma confinement, *US Patent*, 9,959,942, (2018).

[45] T. J. McGuire, System for supporting structures immersed in plasma, *US Patent*, 9,959,941, (2018).

[46] T. H. Rider, A general critique of inertial-electrostatic confinement fusion systems, *Physics of Plasmas* **2**, 6, pp. 1853–1872 (1995).

[47] J. D. Huba, *NRL Plasma Formulary*. Naval Research Laboratory (2013).

[48] E. J. Lerner, S. M. Hassan *et al.*, Confined ion energy 200 keV and increased fusion yield in a DPF with monolithic tungsten electrodes and pre-ionization, *Physics of Plasmas* **24**, 10, p. 102708 (2017).

[49] V. A. Gribkov, A. Banaszak *et al.*, Plasma dynamics in the PF-1000 device under full-scale energy storage: II. fast electron and ion characteristics versus neutron emission parameters and gun optimization perspectives, *Journal of Physics D: Applied Physics* **40**, 12, p. 3592 (2007).

[50] E. J. Lerner, Prospects for $P^{11}B$ fusion with the dense plasma focus: New results, *Current Trends in International Fusion Research*, p. 25 (2008).

[51] J. Slough, D. Kirtley & C. Pihl, Advanced fuel cycle and fusion reactors utilizing the same, Patent PCT/US2015/014904, World Intellectual Property Organization (2015).

[52] U.S. Department of Energy, Office of Health, Safety, and Security & Office of Classification, Restricted Data Declassification Decisions 1946 to the Present, Tech. rep. (2002).

[53] R. J. Goldston, Climate change, Nuclear power, and nuclear proliferation: Magnitude matters, *Science & Global Security* **19**, 2, pp. 130–165 (2011).

[54] R. J. Goldston, L. R. Grisham & G. W. Hammett, Climate change, nuclear proliferation and fusion energy, Tech. rep., IAEA 2010 (2010).

[55] A. Glaser & R. J. Goldston, Proliferation risks of magnetic fusion energy: Clandestine production, covert production and breakout, *Nuclear Fusion* **52**, 4 (2012).

[56] R. J. Goldston & A. Glaser, Safeguard requirements for fusion power plants, Tech. rep., Institute of Nuclear Materials Management (2013).

[57] B. Kuteev, E. Azizov *et al.*, Steady-state operation in compact tokamaks with copper coils, *Nuclear Fusion* **51**, 7, p. 073013 (2011).

[58] S. Dean, Fusion power by magnetic confinement program plan, *Journal of Fusion Energy* **17**, 4, pp. 263–287 (1998).

[59] International Energy Agency, *Detailed country RD&D budgets database*. IEA (2017), wds.iea.org.

[60] A. Rhodes, J. Skea & M. Hannon, The global surge in energy innovation, *Energies* **7**, 9, pp. 5601–5623 (2014).

[61] Frankfurt School of Finance & Management, *Global trends in renewable Energy Investment*. UN Environment's Economy Division and Bloomberg New Energy Finance (2017).

Index

A

agriculturalists
  energy sources, 4
air conditioners, 12
Al-Qaeda, 308
Alcator C-Mod, 231, 280, 301, 361
Alpha Centauri, 348
aneutronic *see* p-B, 292
ARC, 248
Argentina *see* Ronald, 179
Arkhipov, Vasili, 305
arsenals by country
  nuclear weapons, 304
ASDEX-U, 129, 193, 277
Aston, Francis, 177
atmospheric pollution, 238

B

B-59 submarine, 305
bald spot, 103
banana orbits *see* super-bananas, 118, 190
baseload sources, 46
bell curve, 39, 83
Bell Telephone Laboratory, 184
Beria, Lavrenti, 189
beryllium, 162, 224, 324
Bethe, Hans, 178
Big Bang, 12
binding energy, 70–71
biomass, 35

Boeing 747 airplane, 353
Boltzmann constant, 69
bootstrap current, 189
bootstrap multiplication *see* bootstrap current, 254
brains, 59
bravery, 59, 335
brawn, 59
breakeven, 135
Breakthrough Starshot, 353
breeder reactors, 26, 330
  fuelling proliferation, 333
bremsstrahlung *see* p-B, 293
burning plasma *see* ignition, breakeven, triple product, Lawson criteria, 136, 214

C

C-2U, 278
Californium, 316
CANDU reactors, 327
capital cost, 240
carbon capture and storage, 39
Carnot limit, 31
catalyzed D–D fuel cycle, 85, 355
central solenoid, 145, 283
chain reaction, 314
chemical propulsion, 352
Chernobyl, 24
Chicago Pile-1, 311

372 Index

classical transport, 123
climate change, 238–239
CNO cycle *see* stars, Bethe, Hans, 178
cold fusion, 77
Commonwealth Fusion Systems, 301
confinement, 82
    electrostatic, 112
    empirical scaling laws, 251
    energy confinement time, 90
    toroidal magnetic, 109
    volume to surface area ratio argument, 256
conservation of momentum, 349
    frozen lake argument, 350
convective eddies, 32
conventional spacecraft, 350
Coriolis force, 32
cost of electricity, 239
critical mass, 315
cross-section, 77–78
Cuban Missile Crisis, 304
current drive, 144
    electron cyclotron, 152
    electromagnetic wave, 150
    inductive, 145
    neutral beam, 148
cusp geometry *see* Lockheed Martin, 288

D

D–^{3}He fusion, 85, 300, 354
D–D fusion, 20, 85, 202, 301, 354
D–T fusion, 20, 83, 161, 202, 269, 281, 287, 292, 337
Darwin, Charles, 175
Debye length, 99
dense plasma focus *see* Lawrenceville plasma physics, 297
deuterium
    abundance on Earth, 22
diagnostics, 164–165
diffusion *see* random walk diffusion, 123
DIII-D, 277

direct drive *see* indirect drive, inertial confinement fusion, 272
dirty bombs, 321
dispatchable sources, 46
disruptions, 153–154, 222, 244, 267
    mitigating, 155
divertor, 157, 194, 220, 256
double-edged sword
    nuclear energy as blueprint, 345
    nuclear physics, 310
    technologies, 311

E

early hominids, 3
Earth–Moon system, 16, 41
Eddington, Arthur, 178
edge localized modes *see* ELMs, 194
Edison, Thomas, 6
electric field, 94, 150
electromagnetic force, 74
electromagnetic induction, 5, 12
electromagnetic repulsion, 67
electromagnetic waves, 269
electromagnetism, 90
electromagnets, 93
electrons, 65
    in light bulbs, 6
electrostatic, 95
ELMs, 194, 222
ELMO bumpy torus, 114
empirical scaling law, 130, 251
enrichment, 333
energy
    conservation of, 11
    flows of, 13
energy hierarchy, 58
energy storage, 50
Enola Gay, 305
entropy, 11
EPED, 197
exhaust velocity, 350
expanding electrical grids, 54
external power, 249

F

Faraday, Michael, 5
Fat Man, 305
  fissiled percent, 322
fertile material, 330
field-reversed configuration, 290
first wall, 153, 222
fission
  proliferation, 333
fission reactors, 325
  climate versus nuclear security tradeoff, 335
fission–fusion hybrids, 332
flow, 197, 245
Fokker–Planck simulations *see* gyrokinetics, 209
formation of fossil fuels, 37
Forrest, Michael, 188–189
fossil fuels, 37
Fukushima, 24
fusion, 13, 16, 19
  enrichment in fusion blanket, 337
  proliferation, 336
fusion fuels, 83
fusion power density, 243
fusion reactor
  design, 237
  disabling a proliferator, 338
  small fusion system, 339
  smallest planned, 159
  timescale for blanket proliferation ramp-up, 340
fusion thruster, 354

G

gas centrifuges, 318
gaseous diffusion, 317
General Atomics, 198, 279
General Fusion, 284
geopolitics, 343
geothermal, 16, 27
global zero, 343
gravitational confinement, 89
gravity, 13, 89

gravity-assist, 351
Greenwald limit, 251
gun-type bomb, 323
gyrokinetic simulations, 252
gyrokinetics, 206, 252
  scale separation, 207
gyroradius, 92

H

H-mode, 130, 193, 206, 252
hairy ball theorem, 103
Halite-Centurion, 274
half-life, 20
Harwell, 187
heat death of the Universe, 12
heat flux, 221
heating, 14, 218
  electron cyclotron, 152
  ion cyclotron, 152
heavy element synthesis, 15
heavy elements, 16
heavy water reactors, 327
Heisenberg's Uncertainty Principle, 79
Helion Energy, 300
heliotron, 114
helium-3
  abundance, 354
hex, 317
Hiroshima, 322, 336
hohlraum *see* inertial confinement fusion, 272
hydroelectric, 39
hydrogen bomb, 89, 304, 324
hydropower, 17

I

IAEA, 334, 339
ignition, 131, 250, 269
  MCF and ICF ignition differences, 274
implosion bomb, 323
inboard *see* torus terminology, 110

indirect drive *see* direct drive, inertial confinement fusion, 272
inductive heating, 147
Industrial Revolution, 4
inertial confinement fusion, 269
    weaponization propspects, 271
intercontinental ballistic missiles *see* ICBM, 309
intercontinental electrical grids, 55
Intergovernmental Panel on Climate Change, 9
internal transport barriers *see* pedestal, 196
intermittency, 30, 46
INTOR, 225
IPA, 278
Iron Curtain, 181
isotopes, 20, 73
isotope effect, 204
isotopic semantics, 73
ITER, 137, 203, 211, 237, 248, 257, 273, 302, 353, 361
    ignition, 214
    Q, 213
    strategy, 216
Ivy Mike, 304

J

JET, 140, 155, 169, 198, 200, 260, 273, 277
jet engine, 349
JT-60, 260
JT-60SA, 140, 169, 361
JT-60U, 140, 277, 361
Juno spacecraft, 347

K

K-DEMO, 361
Kelvin, Lord, 177
Khrushchev, Nikita, 187
kink limit, 246
Kremlin, 186
Kurchatov, Igor, 188
Kurchatov Institute, 185

L

L-mode, 193, 252
Lamb, Horace, 127
Landau energy levels, 298
Landau damping, 151
Landau, Lev, 151
Langmuir, Irving, 165
Langmuir probes, 165
laser enrichment, 320
lattice structure, 156
Lavrentyev, Oleg, 186
Lawrenceville Plasma Physics, 296
Lawson criterion *see* triple product, 131, 324
Lawson, John, 131
levitated dipole, 114
limiter *see* divertor, 157, 194
linear magnetic, 100
lithium, 21, 242
lithium pebbles, 163
lithium-6, 84, 161
lithium-7, 161
lithium-ion batteries, 22, 51
Little Boy
    fissiled percent, 322
Lockheed Martin, 287
Lufthansa Flight 181, 199

M

magnet(s), 139
    permanent, 93
magnetic confinement fusion, 247
magnetic field, 91, 150, 247
magnetic islands, 117, 265
magnetic mirror, 100, 355
magnetic surfaces, 114, 144, 158, 191, 265
    open versus closed, 158
magnetized target fusion *see* MTF, 284
magnetohydrodynamics *see* MHD, 121
Manhattan Project, 318
Mars, 347
mass–energy equivalence, 77

material survivability, 255
matter–antimatter annihilation, 352
mechanical stress, 142, 283
Mercury, 181
Mercury laser *see* NIF, 276
messy engineering endeavor, 259
MHD, 209, 246
MHD stability, 153
mini-golf, 68
Mini-Sphere, 278
MIT, 301
Model C stellarator, 189
moderator, 326
Moore's Law *see* triple product, 136, 205
MRI machines, 140
MTF, 284, 300
Munich, 199

N

Nagasaki, 305, 322
neoclassical transport, 126, 265
net electric power, 241–242
net electricity, 169
neutral beam, 220
    negative ion acceleration, 220
neutral beams, 291
    efficiency, 149
neutron capture
    cross-section, 331
neutron flux, 156
neutron multiplication, 224
neutron multiplication factor, 162
neutron multipliers, 162
neutron shielding, 292
New York Times, 180
NIF, 273, 336
niobium–titanium, 184
niobium–tin, 184, 248
Nixon, Richard, 224
Nobel Peace Prize, 186
Nobel Prize, 183
non-inductive *see* current drive; neutral beam, 149

non-renewable, 19
North Korea, 339
nucleons, 65
nuclear energy transition, 341
nuclear fission, 16, 23
nuclear proliferation, 238, 359
nuclear potential, 66
nuclear security, 336
nuclear weapon, 303
    boosted implosion bomb, 323
    defenses, 309
    gun-type bomb, 322
    hydrogen bomb, 323
    implosion bomb, 322
    inspectors, 340
    neutron initiator, 324
    proliferation with increased tritium availability, 342
    significant quantity, 334
    tamper, 322
    Teller–Ulam design, 325
    weapon designs, 321
    yield, 303
nuclear winter, 308

O

ocean waves, 44
Onnes, Heike, 181
outboard *see* torus terminology, 110
Oxford, 199

P

p-B fusion, 290
particle drifts, 104
    $E \times B$ drift, 106, 128
    ∇B drift, 106, 194
    curvature drift, 106
Pauli exclusion principle, 183
pedestal, 193
Pelamis, 46
Perhapsatron, 114, 181, 285
photosynthesis, 35
plasma, 86
plasma current, 114, 264

maximization, 246
plasma flow, 197, 245
plasma gain, 169
plasma heating, 144
plasma power engineering multiplication factor, 170
plasma power multiplication factor, 135
plasma pressure, 244
plasma shaping, 198
Pluto, 321, 347
plutonium, 90
    plutonium-239, 25
    production, 320
    reactor-grade, 329
    weapons-grade, 321
poloidal field coils, 144
poloidal *see* torus terminology, 110
polonium-210, 162
power multiplication, 170
Princeton, 189, 198
Princeton University, 180
profits, 238
proliferation, 333
propellant, 349
proton–proton chain *see* CNO cycle, stars, Bethe, Hans, 179
proton–proton fusion, 28
public relations, 359
pure fission weapons, 325

Q

quasineutrality, 97–98, 150

R

radioactive waste, 167
random walk diffusion, 123
rate of energy consumption, 18
Rayleigh–Taylor instability, 271, 285
remote maintenance system, 204
renewable, 19, 57
resistivity, 141
Richter, Ronald, 179, 187
right-hand rule, 93
robotic maintenance, 169, 257
rocket equation, 351
role model
    effect of fusion technology, 344

S

safety factor, 115
Sakharov, Andrei, 186
Saturn, 347
scattering collision, 67
scientific notation, 26
seasonal energy storage, 52
Seebeck effect, 6
seeds, 35
shaping, 198
    D-shape, 198
shattered pellet injection *see* disruptions, 156
significant quantity, 340
solar, 16, 28
Solar System, 17, 348
    energy flows, 15
space capsules *see* divertor, 160
space colonization, 349
SPARC, 301
spent fuel
    current world production, 334
spherical tokamaks, 281
spheromak, 114, 287
Spitzer, Lyman, 113, 180, 263
ST40, 278
Stalin, Joseph, 187
stars, 175
    possible fusion reactions, 178
    red giant phase, 15
steam engine, 4
steam turbine, 6
stellarator(s), 114, 263
    ignited, 267
stochastic regions, 117, 265
strong nuclear force, 66
Sun, 347
    lifetime of, 176
super-bananas *see* banana orbits, 118
super-duper H-mode, 259
supercomputers, 204

superconductivity, 141, 181, 248
    Cooper pairs, 183
    type I, 184
    type II, 184
superconductor, 283
    high-temperature, 185, 283, 301
        REBCO, 186
    materials, 182
    problem with neutrons, 163
supernovae, 14–15
surface-to-air missiles, 309
Sword of Damocles, 306
Symmetric Tokamak, 189

T

T-1, 188
T-3, 113, 189, 263, 278
T-7, 185
T4, 278
TAE Technologies, 280, 290
Tamm, Igor, 186
TCV, 201, 280
technetium, 24
temperature, 83
TFTR, 203, 277
    tritium detection, 339
thermal equilibrium, 82
thermodynamic efficiency, 31
thermonuclear bomb *see*
    nuclear weapons — hydrogen
    bomb, 324
thermotron, 180
Thor, 110
Thomson scattering, 166
thorium, 321
Three Mile Island, 24
tidal, 16, 41
TNT, 304
toast
    making of, 7
tokamak, 113
Tokamak Energy Ltd., 281
Tore Supra, 141
toroidal
    torus terminology, 110
toroidal field, 140

toroidal field coils, 198
toroidal symmetry, 264
torsatron, 114
torus, 104
torus terminology, 110
trapped particles, 118
TRIAM-1M, 185
Trinity, 303, 344
triple product, 135, 267
tritium, 20, 161, 338
    cost, 202
    current reserves, 341
    detecting use in fusion reactor,
        339
    increased availability
        proliferation risk, 342
tritium breeding, 215
tritium breeding blanket, 160, 247
Troyon limit, 243, 282
    game of chicken, 244
    violation, 245
Tsar Bomba, 304
Tuck, James, 181
tungsten, 224
turbulence, 206, 255
turbulent eddy, 128, 197
turbulent transport, 126

U

U.S.S. Beale *see* Cuban Missile
    Crisis, 305
units of energy and power, 18
units of nuclear energy, 69
uranium, 90
    enrichment of U-238, 315
    fission reaction, 312
    uranium deposits, 25
    uranium-235, 23
    uranium-238, 24
US Energy Research and
    Development Administration, 360
US Geological Survey, 21

V

vacuum vessel, 164

W
W7-X, 265, 280
Wagner, Fritz, 193
waste repository, 168
wave, 44
WEST, 141
wind, 17, 30
work, 4
World War II energy release, 304

X
X-point, 158, 194, 256
X-rays, 272

Y
yellowcake, 317
Yeltsin, Boris, 307

Z
ZETA, 165, 187, 297

Printed in Great Britain
by Amazon